Birkhäuser

Abraham A. Fraenkel

This portrait was photographed by Alfred Bernheim, Jerusalem, Israel.

Abraham A. Fraenkel

Recollections of a Jewish Mathematician in Germany

Edited by Jiska Cohen-Mansfield

Translated by Allison Brown

Birkhäuser

Author
Abraham A. Fraenkel (1891–1965)
Jerusalem, Israel

Editor
Jiska Cohen-Mansfield
Jerusalem, Israel

Translated by Allison Brown

ISBN 978-3-319-30845-6 ISBN 978-3-319-30847-0 (eBook)
DOI 10.1007/978-3-319-30847-0

Library of Congress Control Number: 2016943130

Cover illustration: Abraham A. Fraenkel 1939. Drawing by Leo Robitschek, Jerusalem, Israel. Source: Family collection

Printed on acid-free paper

This book is published under the trade name Birkhäuser.
The registered company is Springer International Publishing AG Switzerland (www.birkhauser-science.com).

Acknowledgements

The publication of this translation of Abraham Fraenkel's autobiography was initiated by his son, Benjamin Fraenkel, with support from his siblings, Rahel Bloch, Tirza Cohen, and Aviezri Fraenkel. Indeed, it was Benjamin Fraenkel's last request of me before he passed away that I complete this project. I thank the siblings, as well as Benjamin's wife, Judith Fraenkel, for their trust and support during this process. I wish to express my gratitude to the many people who supported me in the process of preparing this book. Some, including Prof. Moshe David Herr, provided first-hand information about Fraenkel, whereas others, such as Yuval Fraenkel, searched for materials about Fraenkel in the archives of the National Library of Israel. Sharon Horowitz, Bina Juravel, and others looked for bibliographic details. Several individuals helped decipher the meaning of specific texts in the book, including Prof. Deborah Gera, who helped with the translation from the Greek, Avraham Fraenkel (son of Jonah), who assisted in figuring out the analysis of the *piyyut*, and Prof. Jonathan Rosenberg, who checked the translation of the mathematical portions of the book, as well as some others. Still others helped with the many steps needed to bring this book to fruition, including Anne Birkenhauer, Mimi Feuerstein, Michael Fraenkel, Dina Goldschmidt, David Koral, Prof. Jerry Muller, Rabbi Dr. Isaac Sassoon, Rabbi Michael Swirsky, Pnina Wandel, and others. My husband, Allen Mansfield, and my three children, Jonathan, Hillel, and Ariella, all helped with different aspects of the book preparation, as did my sisters, Noah Liel and Orina Cohen, and my brother-in-law, Yair Liel. I am indebted to Prof. Magidor for writing the current introduction to the book. Finally, I thank the translator, Allison Brown, who, besides translating, thoroughly researched the background of the book, and the editor, Susan Kennedy, who made the book more accessible to readers.

The autobiography was originally published with support from the Leo Baeck Institute Jerusalem. In the current edition, I would like to thank my contacts at Springer Publishing, Anna Maetzener and Sarah Goob. Ms. Maetzener suggested that we augment the original book with a new introduction by a current prominent mathematician, as well as a bibliography, photographs, and a chapter about Fraenkel's life after the events described in the original volume. All these, as well as family trees, have been added to this volume. This translation also includes new footnotes in which the translator and/or I clarified points in the text. These new footnotes are indicated with two asterisks in order to distinguish them from the original footnotes.

Jerusalem 2015 Jiska Cohen-Mansfield

Contents

Foreword to the 2016 English Edition

Professor Abraham HaLevi Fraenkel was my mathematical grandfather, that is to say the teacher of my teacher Professor Azriel Levy. The interaction I had with him was rather minimal: When I started my undergraduate studies at the Hebrew University of Jerusalem in 1962, Fraenkel was already emeritus and he passed away before I got my bachelor's degree. I did attend a course on the Jewish calendar that he taught as professor emeritus, but I was too shy to have any meaningful interaction with him. This being said, Fraenkel had a very profound impact on my career choices. In fact, he is indirectly responsible for my becoming a mathematician and especially for my interest in Set Theory.

I was 13 years old when, browsing through the books in a used bookstore in Netanya, the district town of the area in which I grew up, I ran into a series of four thin paperback volumes with the Hebrew title *Mavo Le-Mathematica* (Introduction to Mathematics). At that point, I hardly had any idea what "mathematics" was. My elementary school education in mathematics was limited to rather technical routine and boring arithmetical procedures. I started leafing through these books and randomly reading passages. Within a few minutes, it was clear to me, in spite of the fact that I did not fully understand what I was reading, that I was facing a building, very abstract but of sublime beauty. I fell in love with it and, right there on the spot, decided that the study of the architecture of that building would be the main theme of my life. I purchased the books and they still constitute the cornerstone of my mathematical library.

Fraenkel wrote that series of books over a period of several years from 1938 till 1945, but because of technical difficulties arising from the Second World War and Israel's War of Independence, their publication by the Hebrew University Press was delayed and spanned 15 years, from 1942 until 1957. Even from the perspective of more than half a century, I still consider this book to be by far the best of its kind. The volumes cover most of the important basic concepts of modern mathematics. Naturally, since it is Fraenkel, there is an excellent volume on the basics of Set Theory. (My fascination with the exposition in this volume is responsible for the fact that most of my mathematical work is in Set Theory). Besides its wide coverage, it is unique in the connections it makes between developments in mathematics and the general culture, especially philosophy. Fraenkel was very careful in choosing the book's subjects, which he describes in full technical detail, but most importantly, there is a unique conceptual clarity to the basic notions and the motivations for their introduction.

In a deep sense, the conceptual clarity, the ability to see the essential features of the issues, and the succinct formulations that emerge from them are characteristic of Fraenkel's mathematical contributions. His two most famous contributions—the addition of the axiom of replacement to the standard axiom system of Set Theory, and the method of showing the independence of the axiom of choice from the Set Theory containing atoms—are not characterized by very elaborate technical developments, but are a major breakthrough in conceptual clarification, of finding the right formal explication of a notion that previously existed only intuitively and vaguely. His Ph.D. thesis on the p-adic numbers, essentially one of the first works introducing the important concept of "ring", has the same character of giving rigorous definition to concepts that were formulated only vaguely by his thesis supervisor, Kurt Hensel. The same clarity and the same ability to see the essential issues in a murky social or political situation are evident in Fraenkel's autobiography, even when he deals with domains that are very far from mathematical.

There are three major themes in this autobiography which have relevance to present-day contexts and can be examined from a contemporary perspective. The story of the book is mainly an account of an individual Jewish-German family, but the broader context is the last generation of Jewish-German society before its demise in the Second World War. More specifically, it is a description of the challenges facing an orthodox conservative minority that only a few generations prior to the described events had been enclosed within almost impenetrable physical and social boundaries, but which, when opportunities for integrating and operating in society at large opened up, at least formally, found itself faced with the tension between the desire to retain a traditional lifestyle and the pressure to integrate, or even assimilate, into general society. A variation on these challenges still exists today for traditional minorities in many developed countries. Similar dilemmas are currently faced by the Haredi community (strict orthodox Jews) in Israel, for instance, whether a core curriculum of general subjects ("*Limudei-liba*") should be included in the Haredi educational system.

The social and the cultural milieu of the several generations of Fraenkels described here is a very finely nuanced balance between strictly observant orthodoxy and a very active and eager participation in the general academic, political, and cultural environment. A very telling detail is the friendship that the strictly orthodox Fraenkel had with Christian theologians. Of course, this participation came to an abrupt end with the rise of the Nazi regime. An interesting twist is the critical attitude of major parts of the orthodox community to Fraenkel's Zionist involvement, including criticism of his acceptance of a position at the Hebrew University. The possibility of such a balance is definitely a lesson to be learned, in particular in present-day Israel.

The second theme, which requires more extensive commentary from a contemporary point of view, is the role played by Fraenkel's mathematical contributions in present-day Set Theory. Set Theory was created (or discovered—the right term is a matter of philosophical conviction) by George Cantor in the last decades of the nineteenth century. Cantor's version of Set Theory was to a large extent naive and intuitive. Especially naive was Cantor's unrestricted use of the principle of comprehension, according to which for every condition $\Phi(x)$ there exists the set that contains exactly those objects x which fulfil this condition. Towards the end of the nineteenth century, Cantor and others (e.g. Burali-Fori) became aware that such unrestricted use of the principle of comprehension leads to inconsistencies in Set Theory. The ultimate antinomy was discovered by Bertrand Russell (the famous "Russell's paradox") in 1901 when one applies the principle of comprehension to the property "x is a set that is not a member of itself". These antinomies created a crisis that threatened the very foundation of the theory and raised serious issues about the foundation of mathematics. A way out was suggested in 1908 by Zermelo who restricted the principle of comprehension by applying it only to the collection of elements that are already included in a given set. (The modified principle is called "the axiom of separation".) This requires also formulating a list of natural principles stipulating the existence of certain sets to which the restricted principle of comprehension can be applied.

Zermelo's system of principles (or "axioms") seemed to provide a sound basis for Set Theory, hopefully without including a contradiction. However, there was still a vagueness in Zermelo's formulation of the principle of separation (translated from the German):

> If the statement $\mathrm{U}(x)$ is definite for all members of the set M, then the set M has always a subset M_{U} which contains those members of M for which $\mathrm{U}(x)$ is true and only those members.[1]

The problem with this formulation is the vagueness of notions like "statement" or "definite".

Fraenkel in 1922 gave an explicit formulation of these concepts by specifying a class of functions defined by combinations of functions introduced by the other Zermelo axioms, and interpreting "definite statement" as statements of the form $f(x) \in g(x)$ or $f(x) \notin g(x)$ where the functions f, g are in the class. An equivalent formulation was given independently a year later by Skolem. Furthermore, Fraenkel realized that there was a natural axiom that was missing from Zermelo's axioms, which is implicitly used in many natural constructions. He formulated this axiom which he called "the axiom of replacement". The Zermelo axiom system as modified and augmented by Fraenkel became known as the Zermelo–Fraenkel axiom system or ZF. (When it includes the axiom of choice, it is denoted by ZFC.). ZFC very quickly became the canonical axiom system in which Set Theory is formalized. One reason for its almost universal acceptance was the fact that the

[1]Quoted in *Foundation of Set Theory* by A. Fraenkel, Y. Bar-Hillel, and A. Levy, 2nd edition, North Holland 1973, page 36.

axioms seemed to be very natural and such that they fit very well the intuitive notion of the concept of "set". Apparently, ZFC is a framework that is capable of including all of mathematics, and it seems to be free of the contradictions that afflicted the naive Cantorian Set Theory.

Set Theory, in the Zermelo–Fraenkel formulation, turned out to be a very fruitful mathematical theory, but many fundamental open problems persisted for a long time. The best-known open problem was the Continuum Hypothesis, which goes back to Cantor. In fact, it was the first problem on the list of central problems of mathematics presented by Hilbert at the Second International Congress of Mathematics in Paris in 1900. In 1938, Kurt Gödel proved that the Continuum Hypothesis could not be refuted in Set Theory. It still did not settle whether the Continuum Hypothesis was derivable in ZFC.

The next major breakthrough of Set Theory occurred in 1963 (two years before Fraenkel passed away) when Paul J. Cohen invented a technique (termed "the method of forcing") for constructing models of ZFC with varying properties. In particular, he constructed a model in which the Continuum Hypothesis failed. Hence, this central problem of Set Theory could not be decided on the basis of ZFC. The Continuum problem was not unique. Using the forcing method, many open problems and other mathematical fields, like Analysis, Algebra, and Topology, were shown to be undecided on the basis of ZFC.

In some sense, the phenomenon of independence was not unexpected. The famous theorem of Gödel (1931), known as the incompleteness theorem, claims that any mathematical theory rich enough to express some basic arithmetic facts (ZFC is definitely rich enough in this sense) is incomplete. Namely, it contains a statement which cannot be decided on the basis of the given theory. The surprise lay in the fact that the independent problems were not artificially constructed problems, but problems central to the field. This raised a deep philosophical problem: What is the meaning of independence? How do we settle the undecided problem? Is there a definite answer to the problem or does independence mean that the mathematical objects do not have an objective absolute existence? If, as many mathematicians believe, the mathematical objects represent an objective absolute reality of some kind, the way to get additional information about this reality and settle the independent problems is by studying extensions of ZFC.

Finding natural extensions of ZFC that would settle many of the undecided problems became a central research programme in contemporary Set Theory. These attempts assumed several directions like strong axioms of infinity (assuming that there are larger and larger sets), forcing axioms (intuitively meaning that a set whose existence can be imagined does exist) or canonical inner models. While there were several success stories where large classes of independent problems were settled by such extensions, none of the extensions of the Zermelo–Fraenkel Set Theory was able to gain the almost universal acceptance of the canonical natural assumptions about the universe of sets that ZFC did.

Fraenkel's second major contribution to Set Theory was also a source of many later developments. It concerns one of the axioms introduced by Zermelo: the axiom of choice (AC) in order to justify several natural constructions. Its introduction as an axiom was initially controversial because of its non-constructive character. But now it is almost universally accepted. An interesting problem was whether the introduction of AC into the axiom system was not redundant, namely did it follow from the other axioms. Fraenkel in 1922 devised a method for showing the independence of the axiom. It did not apply to the accepted version of ZFC but a somewhat different version of Set Theory in which the universe of sets is constructed on the basis of an initial set of "atoms". Fraenkel's method started from a universe of sets with infinite sets of atoms and defined a subuniverse of sets that were invariant under some permutations of the atoms. The method was extended by the Polish mathematician Mostowski. The method is thus known as the Fraenkel–Mostowski method and was used to show the independence of many statements that follow from the axiom of choice. All these applications were for the version of ZFC with atoms.

Since the accepted version of ZFC is without atoms, this work left open the status of AC with respect to the atomless version of ZFC. The forcing method of Cohen once again came to the rescue. Part of Cohen's seminal work was to show the independence of AC with respect to the atomless ZFC. An interesting feature of Cohen's proof is that it has a clear affinity to the Fraenkel–Mostowski method. In fact, Cohen himself in his book about his method[2] points to this affinity. Cohen's work on the independence of AC was followed by a series of results which directly converted results obtained by the Fraenkel–Mostowski method, using forcing to get independent results also for the atomless version of ZFC.

The third theme that is worth commenting on is the very fundamental role of Fraenkel in the formation of the Hebrew University of Jerusalem. This aspect of Fraenkel's activity is represented only to a very limited extent in this volume because the time span described here concludes with Fraenkel's joining the Hebrew University and settling in Jerusalem. The next period, in which he made his most significant contributions to the university, was supposed to be covered in the subsequent volume of this autobiography. Sadly, Fraenkel passed away soon after the conclusion of this volume, so we do not have his version of his intensive activity in a leadership role at the young university.

Plans for establishing the Hebrew University of Jerusalem as part of the Zionist venture of recreating the Jewish commonwealth in Palestine had been taking shape since the beginning of the twentieth century. They become much more concrete after the First World War and the establishment of the British mandate for Palestine. The cornerstone for the campus was laid in 1918 and the opening ceremony took place in 1925. From the early stages, the character of the budding institution was a subject of great controversy. Many of the leaders of the university emphasized their ambition to create a research university of world caliber. (One needs to appreciate the boldness, or better the impertinence, of such a vision in view of the poor conditions and scant physical and academic resources available in Jerusalem in the 1920s.) Others, however, sought to establish a teaching institution whose

[2]*Set Theory and the Continuum Hypothesis*, P.J. Cohen, Benjamin 1966.

main mission would be to serve the needs of the small Jewish community in the country at the time, or offer the option of a college education for the many Jewish students from Eastern Europe, for whom admittance to their local universities was limited due to discriminatory policies. A similar dividing line existed between people who wanted the university to compete on the world scene and therefore to concentrate its research activity on subjects of universal interest as opposed to those who wanted the university to primarily serve the immediate needs of the country.

Prominent among the supporters of the first view was Albert Einstein, who had been involved since 1919 in the attempts to create the university. The other side can probably best be identified with Zionist leader Ze'ev Jabotinsky. The Zionist leader Chaim Weizmann can probably be described as the man in the middle. This controversy was bound up with many personal issues, severe criticism of the way Magnes was administering the university, and even external political preferences. Confrontations became acute in the late 1920s and the early 1930s. It reached the point where Einstein was so frustrated with Magnes that he withdrew his engagement with the university. This was a very serious blow to the project, since the involvement of a world academic leader like Einstein was one of the main assets of the young university.

This is precisely the period when Fraenkel joined the university. In retrospect, this move had a very deep impact on the future course of the institution. The greatest challenge that faced the new university was the recruitment of new faculty. This was an especially daunting task given the great chasm between the declared ambition of many of the founders of the university of creating a world-class research institution and the poor conditions in Jerusalem of the 1920s. The chances of attracting established academics to Jerusalem seemed rather slim. In fact, many members of the new faculty of the university in its initial years were rather young and junior. (It should be said, however, that many of them developed to become world-class scholars.) Fraenkel was one of very few who joined the young university, out of deep commitment to the Zionist idea, who had an established academic status as a senior scholar. In fact, one can claim that the external academic status of Fraenkel was much higher than that of any of the faculty of the university in the early 1930s. It is true that Fraenkel replaced another very distinguished mathematician, Edmond Landau of Götingen, who very seriously considered joining the university. He spent the winter term of 1928 in Jerusalem, but personal conflicts with the then chancellor of the university, Magnes, caused him to drop his plans to settle in Jerusalem. So when Fraenkel moved to Jerusalem in 1929, he was the best-known scholar among the small group of faculty members in Jerusalem. No other professor at the university besides Fraenkel had the same caliber of contacts with world-class academics like Einstein or Hilbert. If the ambition of the young university was to become a world-class institution, then recruiting academics of the class of Fraenkel was a necessary condition.

It is no coincidence that Fraenkel was Einstein's main contact in the university and his main source of information about it. When Einstein threatened to cut his involvement in the university, Fraenkel played a central role in convincing him to stay involved. Fraenkel's leadership position was enhanced substantially when, following the pressure of Einstein and Weizmann, the governance structure of the university was reformed. The role of the chancellor was limited to external representation of the university (the title of the position was changed to "president"), and in parallel a position of academic head of the university, called "rector", was created. Fraenkel served as the second rector of the university and as such played a major role in shaping the academic future of the institution.

His position on the controversial issue of academic policy was clearly to put a great emphasis on the research excellence of the university and its international status, even if it meant a certain preference for issues and subjects which were of universal interest over issues of local interest. The fact that the Hebrew University in particular and Israeli science in general have an excellent world reputation is due to a large extent to this stance. This is especially evident in the field of pure mathematics, where Fraenkel's influence is directly felt.

This does not mean that Fraenkel was a typical ivory tower professor. He was committed to the role of the university in society at large. He played a very active role in the educational system. From 1933 until early 1950, Fraenkel chaired the university committee on high schools, which influenced the curricula and the pedagogic methodologies of the Hebrew high schools. He had a special interest in adult education. Besides his role as the chair of the committee on popular education, he spent enormous time and effort in delivering popular lectures on advanced topics in mathematics all over the country. As he mentions in this volume, he reached the remotest corners of the country, sometimes on horseback or donkey. His popular book *Mavo Le-Mathematica*, which, as mentioned above, had such a deep impact on my career, was written with the same goal of bridging the gap between pure research, which is the main role of the university, and general society. Fraenkel was a fine example of a balance between academic commitment to pure scholarship in the pursuit of knowledge for its own sake and social commitment.

Professor Abraham HaLevi Fraenkel was positioned at critical junctures of vastly different domains: The last decades of the German-Jewish community before its demise, the establishment of a firm foundation for Set Theory and Mathematics, and the formation of the new Jewish commonwealth in Israel, especially its academic and scientific infrastructure. In all these domains, he combined a conceptual clarity, deep knowledge of the relevant issues, and ideological commitment. This autobiography is a fascinating and illuminating testimony of a unique individual who was both an important player in and a keen observer of these different junctures.

The Hebrew University
Jerusalem, Israel
February 2015

Menachem Magidor

Foreword to the 1967 German Edition

Professor Abraham (Adolf) Fraenkel did not live to see the publication of this book. Early in the morning of October 15, 1965, the spry 74-year-old went for his daily swim in Jerusalem, where he had been living for almost four decades. A few hours later his heart stopped.

He was ready. He often spoke about death during his last few months. He believed he had completed his lifework and that it was well done. His autobiography, the first part of which is this book, was meant to be the "final chapter" of his complete works. He continued his scientific work until the end: six months before his death, he gave me his revised section for the forthcoming second edition of our joint book *Foundations of Set Theory*.[3] However, he was well aware that, in terms of his creativity, he was long past his prime. As he often mentioned in all seriousness, but with good spirit, mathematicians generally accomplish their greatest work before they turn 30. Indeed, he made his most outstanding contribution to mathematics very early on, at the age of 28, with his fundamental book, *Einleitung in die Mengenlehre* (Introduction to Set Theory), published in 1919.

While mathematical work and research were important to him, the scope of his life and work was much broader. Raised in an Orthodox household, deeply steeped in Jewish tradition, and a Zionist since adolescence, he also contributed towards creating a healthy, viable basis for a Jewish homeland, first in Palestine[4] and, later, in the State of Israel. He considered a good education to be the essential prerequisite for this, not only for youth but also for adults, in schools and at universities, not only in the cities but also in the countryside, and in the most remote kibbutz.

---

[3] The second (revised) edition, by A. A. Fraenkel, Y. Bar-Hillel, and A. Levy, *Foundations of Set Theory* (*Studies in Logic and the Foundations of Mathematics*, vol. 67) (Amsterdam: North-Holland/Elsevier, 1966). There was also a 1973 edition with the collaboration of Dirk van Dalen.**

[4] During the Ottoman Empire and the British Mandate, pre-state Israel was known as Palestine or Palestine/Land of Israel. For the sake of clarity, this book will use the formulation "Palestine/Land of Israel" to denote the region, and "*yishuv*" (Hebrew "settlement") to refer to the Jewish community there from 1860 to 1948, prior to the founding of the State of Israel.**

With unlimited willpower and iron self-discipline, he followed his schedule down to the last detail. This allowed him not only to carry out his teaching and research activities, but also to travel all over Israel, even to the most distant parts, from Dan to Eilat, to inspect middle and upper secondary schools and draft curricula, as well as to give hundreds of lectures. In addition, he also went swimming, hiking, and mountain climbing.

Until he became professor emeritus, Professor Fraenkel, together with Professor Michael Fekete, directed the Institute of Mathematics at the Hebrew University of Jerusalem. The international renown of the Hebrew University in the areas of mathematical logic, abstract set theory, and the foundations of mathematics can undoubtedly be largely attributed to his efforts. His students, Haim Gaifmann, Azriel Levy, Michael Rabin, Eliyahu Shamir, and Abraham Robinson, are among the best in their fields, and, with the exception of Robinson, all teach at the Hebrew University.

In 1938, Professor Fraenkel became the second rector of the Hebrew University, an office he held until the end of 1940. Afterwards, he spent many years as a member of the Administrative Council, chaired numerous commissions, and made his analytical skills and extensive expertise available in many aspects of academic life and research. His concerns were by no means limited to purely academic matters. For almost two decades, he chaired the Hebrew University athletic commission, organizing and, until an advanced age, also often participating in hikes, athletic events, and races.

As part of his great, sustained commitment to education, Professor Fraenkel founded, with others, the Center for Adult Education of the Hebrew University, where he served as director for many years. His intense commitment was also manifest in his unremitting readiness to give popular lectures to various groups on the foundations of mathematics, modern physics, and the Jewish calendar. He spared no effort to reach remote locations, on occasion even riding a donkey on the last leg of an arduous journey.

These activities, by no means independent of one another, were part of his lifelong effort to put into practice Orthodox German Jewry's *Torah im Derekh Eretz* (combining Torah [Jewish religious principles] with proper behavior in civic life). He did this himself and hoped thus to serve as a model to others.

While he was a profoundly religious man, Professor Fraenkel also showed true tolerance. Many of his students, including myself, were not religious, but this never affected his attitude towards us. He firmly believed that religion and science were two separate domains that should not be intertwined. The physical worldview was to be based on purely scientific findings. He felt that it was not rational to try to convert non-believers into believers. Attempts to use political power for religious coercion, as by Israel's National Religious Party, were repugnant to him. This is one of the reasons why he never joined that party.

His dream was to see the entire Jewish people united and unified in Israel. For this reason, he often prayed in Yemenite and other Oriental[5] synagogues, where he

[5]Oriental Jews are Jews of North African and Middle Eastern origin.**

was welcomed and honored. Professor Fraenkel greatly enjoyed taking guests on a personal Friday evening synagogue tour whenever he could. With skill and knowledge, he would explain the particularities of various customs and traditions, down to the smallest detail. Those tours gave many non-Jews, as well as quite a number of Jews, their first objective and vivid picture of these "exotic" congregations. Afterwards, they would be invited to a Sabbath dinner at his home, featuring an Israeli version of Bavarian Jewish custom.

Professor Fraenkel's most lasting impression was undoubtedly left on his own students. In describing what he meant to them, the word "teacher" is entirely inadequate. As an inspector of middle schools and academic secondary schools, he was always on the lookout for mathematically gifted children. He discovered several child prodigies, whom he supported as best he could. To spur interest in mathematics, he wrote a book, *Introduction to Mathematics*, in Hebrew, in which his clear and exciting presentation gave many pupils their first insights into the field. He also contributed significantly to developing an appropriate Hebrew terminology for higher mathematics.

As tolerant as Professor Fraenkel was in general, there were two traits that he could not and would not abide among his students—and not only them: a lack of punctuality and incompetence. Few things aggravated him more than students attempting to cloak incompetence and insufficient understanding in rhetoric, platitudes, or vague formulations. He would take such students to task in a way they would never forget.

Gifted students were assured of his personal favour. The distinguished accomplishments of so many of them were clearly not only due to their scientific aptitude but equally to their teacher's interest and consistent support.

To illustrate this, let me share a personal experience. In the spring of 1937, at one of Professor Fraenkel's seminars on the foundations of mathematics, I gave a talk on logical and semantic antinomies. He liked it and, although he was the veritable master of basic research in mathematics, immediately suggested that I, a 22-year-old novice at the time, expand my talk into a joint project with him, which was indeed published in a French journal two years later.[6] The translator, a French-Jewish mathematician, was unfortunately killed a short time later by the Nazis. That was my first international publication.

What attracted me to Professor Fraenkel was not only his personal interest in my growth and progress and his harsh rejection of all non-scientific metaphysics but also his own farsightedness. He was never just a mathematician. While he paid rigorous attention to the philosophical and logical foundations of mathematics, he did not confine himself to them. He tended towards Platonism as a philosophy of mathematics, namely that mathematical entities fully exist as abstract objects, even at times when this view was not very popular. However, he also gave the best and clearest interpretation of intuitionist views, which he personally did not support.

---

[6] Fraenkel, A. A., and J. Bar-Hillel (1939). "Le Problème des antinomies et ses développements récents." *Revue de Métaphysique et de Morale* 46:225–242.**

Although I did not share Professor Fraenkel's Platonism, favoring instead an ontology-free philosophy of mathematics, this did not mar our student–teacher relationship, or our later collaboration, in the least. He was fully aware that it was impossible to prove Platonism to be the only tenable mathematical philosophy. This view appealed to him personally, and he managed to weather the various foundational crises rather well.

Professor Fraenkel's knowledge of philosophy went far beyond the philosophy of mathematics. During the 1920s, he kept in personal contact with many neo-Kantians and phenomenologists in Germany. However, he never really warmed up to their approaches, probably because their statements about mathematics seemed too unclear and irresponsible.

Thus, Professor Abraham Fraenkel will be remembered as a great mathematician, for whom nothing human was foreign; a strict teacher who did not tolerate superficial knowledge, but warmly supported genuine talent; a *Talmid Chacham* (Jewish scholar), a true student of sages, always strict with himself in religious matters, but tolerant towards others; a Zionist, who held the education of the youth in Israel and the unity of the Jewish people close to his heart, every day of his life; and, while superficially austere and pedantic, this did not conceal his warm personality. How this personality was shaped, and then changed when as a German Jew he returned to Israel, making his way from Munich to Jerusalem, is depicted in the autobiography he left us.

The Hebrew University
Jerusalem, Israel
April 1967

Y. Bar-Hillel

Chapter 1
My Ancestors

Most people have eight great-grandparents, namely, the parents of their four grandparents. While that is the case for my parents, wife, children, and grandchildren, my siblings and I have only six great-grandparents. My father's father, Wilhelm Fraenkel,[1] and my mother's mother, Rosa Neuburger, née Fraenkel, were siblings, so their parents, Abraham and Nanette Fraenkel, count twice. Indeed, the oil-painting portraits I have of them [see below] show a remarkable resemblance to two of my own children.

So, in addition to Abraham Fraenkel, I had only two other great-grandfathers: Benjamin Hirsch Auerbach, father of Rahel Fraenkel, my paternal grandmother;

[1]See the Fraenkel family tree p. 204.**

A.A. Fraenkel, *Recollections of a Jewish Mathematician in Germany*,
DOI 10.1007/978-3-319-30847-0_1

and Joël Neuburger, father of Isidor Neuburger, my maternal grandfather. I have nothing notable to report about Joël Neuburger and his family, who lived in Fürth, Bavaria. Benjamin Hirsch (Zvi Benjamin) Auerbach, born on June 21, 1808, in Neuwied, however, was all the more important. The eldest of 16 children of Abraham Auerbach (1763–1845) and his wife Ester Rebecca, née Oppenheim (1785–1864), he received his Ph.D. in philosophy and Semitic languages at the University of Marburg an der Lahn. In 1837, he married Lea Fraenkel (b. Witzenhausen 1814, d. Halberstadt 1884), daughter of Daubchen and Isaak Eisenmann Bodenheim-Fraenkel. Benjamin Hirsch Auerbach (d. 1872) and his wife Lea had seven children.

In the nineteenth century, Benjamin Hirsch Auerbach, the rabbi in Halberstadt, was one of Germany's leading rabbis. Together with Samson Raphael Hirsch and Azriel Hildesheimer, he was among the founders of "neo-Orthodoxy." He became renowned for several writings displaying his profound Talmudic scholarship, especially *Nachal Eshkol*, a commentary on Rabbi Abraham ben Isaac of Narbonne's *Sefer ha-Eshkol*. There is no need to write about him and his numerous descendants, since their monumental family tree is available in *The Auerbach Family*,[2] published by Siegfried M. Auerbach.

I have three authentic sources of information on my great-grandparents who were most significant from a genetic point of view, Abraham Fraenkel (1792–1858)[3] and his wife Perl Nanette, née Neubauer (1808–1881). The first source is Abraham Fraenkel's Hebrew entries in the first volume (*Bereshit,* Genesis) of the splendidly printed *Chumash Derekh Selulah* (which includes a Judeo-German—i.e., Western Yiddish—translation and commentaries), which was published between 1801 and 1803 in Fürth. These entries, covering the period from 1827 to 1843, refer to the birth of his seven children. The second source is his last will and testament, with the heading "Memorandum of Abraham Fraenkel (sic!), addressed during his lifetime to his four beloved children,[4] Sigmund, Wolf, Jacob, and Rosa, written in Munich on February 3, 1857." It includes 40 folio pages in German, followed by the remark, "I would like to write a conclusion to my memorandum, which is intended solely as a recapitulation of all the aforementioned, in the Jewish national script, which I praised above, and is very precious and dear to me." This conclusion consists of nine pages in Judeo-German. Both parts contain many Biblical verses as well as quotations from the *Aggadah*, all magnificently written in square Hebrew script with *nikkud* (vowels). The entire will is in excellent condition and includes an admonition, to all his children as a group and to

[2]London: Perry Press, 1957.

[3]There are no clear-cut conventions for spelling the name: Fränkel, Fraenkel, [and even Fraenckel, in a few cases]. By and large, though not consistently, the first one is considered the earlier version, roughly up to my parents' generation. For the sake of clarity, Fraenkel is the form used in this book.

[4]Three of his seven children died young.**

each one individually, to preserve the Jewish tradition and adhere to religious law. The last page states the following[5]:

תקריבו אלי כלכם כאיש אחד, לא במעשה אלא ברוחי ובמחשבתי, ליתן לכם את
ברכתי, ברכת אברהם אביכם, לכם ולזרעיכם אחריכם, כולן על הברכה תעמדון,
והמקום יהיה עד ישקיף וירא ה' מן השמים ויענה ויאמר אמן, אמן. אמן על ההווה
ואמן על העתיד... ואסיים במה שהתחלתי, ליתן שבח והודיה לאל, יוצרי, עושי
וקוני. על כל הטובות והחסדים אשר עשית עמדי מהיום היותי על האדמה עד היום
הזה. בידך אפקיד רוחי פדיתה אותי ה' אל אמת, בטוב אלין אקיץ ברחמים טובים. אמן
יום ג' טית שבט תרי"ז פה עיר מינכען הבירה. נאם הקטן אברהם פרענקעל הלוי.

Attached to the will is a draft text for his gravestone, which reads as follows:

פ"נ
העפר דאברהם פרענקעל הלוי מקורו מק"ק פיורדא
תנצ"בה[6]
Abraham Fraenkel

born in Fürth on November 27, 1792, and died on [...], 18..., was a teacher, *shochet* [ritual slaughterer], and cantor in Hofheim, near Hattersheim, in the duchy of Nassau, and then cantor, secretary, and, later, member of the administration, executive committee, and management of three charitable organizations in the Jewish community, and also a wholesaler in the royal capital of

MUNICH
God putteth down the high, and lifteth up the low.
Peace to his ashes
(by his own decree, in his lifetime)

The third source is my mother, who at 97 is mentally still fully alert, here in Jerusalem in June 1965,[7] who told me some things about Nanette Fraenkel, her grandmother. For the first 13 years of my mother's life, starting in 1868, they lived together in the home of her mother, Rosa Neuburger (Nanette's daughter).[8]

---

[5] "Come all of you, approach me together as one person, not in deed, but in my spirit and in my thought. To give you my blessing, the blessing of Abraham your father, for you and your offspring after you. All of you shall stand by the blessing. God will be a witness and shall look down from heaven and respond and say Amen, Amen. Amen for the present and Amen for the future ... and I shall end as I have started, giving praise and gratitude unto God, my creator and maker, for all the good and mercy Thou hast done unto me from when I came upon the earth until this day. Into Thine hand I commit my spirit, Thou hast redeemed me, O Lord, God of truth. May I sleep with goodness and awaken with mercy. Amen. Completed on Tuesday, the ninth day of the Hebrew month of Shevat, 5,617 (1857) in the capital city of Munich. The word of the insignificant, Abraham HaLevi Fraenkel."**

[6] Here is buried the dust of Abraham HaLevi Fraenkel, of the holy community of Fürth, May his soul be bound in the bonds of life.**

[7] Charlotte (Chaya Sara) Fraenkel, née Neuburger, died on October 24, 1965.**

[8] In addition to these authentic sources, there are also some less reliable data from people who did genealogical research on our family.

Rabbi Sigbert Neufeld asserted in an essay "*Vom Ries gauüber Wien nach Elbing*," ("From Riesgau to Vienna to Elbing"), in *Das Neue Israel* 14, (Zurich) 1962, that our Fraenkel family came from the Jewish community in Riesgau, on the border between Franconia and Swabia, i.e., from Oettingen, Wallerstein, Spielberg, Harburg, and other towns. According to the same source,

Abraham Fraenkel came from Fürth, a long-established community near Nuremberg. Indeed, Munich's Jewish community, not founded until the early nineteenth century, recruited their most esteemed early members from Fürth. Suffice it to mention the family of Seligmann Feuchtwanger (1786–1852), who was born and died in Fürth. His four sons, Jacob Loew, Elkan, Moritz, and David, settled in Munich, and, together with Abraham Fraenkel and his descendants, formed the core of the Orthodox Jewish community.[9] For the services in their congregation, the printed edition of the Fürth *minhag* (custom) book was authoritative. Jacob Loew (1821–1890) was cofounder of the still-existing J. L. Feuchtwanger Bank.[10]

Characteristic of the legal situation of Jews in Munich in the first third of the nineteenth century is the fact that Abraham and Nanette Fraenkel's wedding in 1826 took place outside Munich, specifically in Kriegshaber, near Augsburg. This was because of the Bavarian edict of 1813, which decreed that "the number of Jewish families in communities where they presently reside should not be increased, but ... gradually decreased." This edict was not permanently abolished until 1861.

Abraham Fraenkel's childhood in Fürth, Hofheim, and, then, in Munich, must have been very modest and subject to religious and other challenges. At the beginning of his last will and testament, he thanked God, "who lifted me from the lowest position to highly honored positions; from servant of the congregation ... to a congregational leader ..., from a recipient of donations to a wealthy distributor of alms, and, thus, from a tolerated stranger to a citizen and wholesaler in this city." His position as *chazan* (Jewish prayer leader) is verified by his mahzor (prayer book for the holidays), which I possess. It is an excellent, first edition (from 1800) of Wolf Heidenheim of Rödelheim's mahzor, which later circulated in many editions.

---

the family assumed the name Fraenkel, derived from their place of origin Franconia (Franken, in German), while other families took the names Riess or Riesser, Oettingen, and Wallerstein, for similar reasons. David Fraenkel, who served as a rabbi in Dessau and Berlin and Moses Mendelssohn's teacher, and Councilor of Commerce Jonas Fraenckel, the benefactor of Breslau's Jewish Theological Seminary, are assumed to have been among them.

In addition, Dr. Siegfried Asher of Haifa and Dr. Yomtov Bato of Tel Aviv were also interested in the background of Abraham Fraenkel of Fürth-Munich, partly because of his Levitical descent. Traces led to Vienna, specifically to Rabbi Israel ben Koppel Halevi Fraenkel, born there around 1640. After the expulsion of Jews from Vienna, in 1670, he was active in several places, with his last post as the district rabbi in Würzburg, which he held until his death in 1700. The name Koppel also came up in our family, as Abraham Fraenkel's brother and a son had this name. The essay by Dr. Yomtov Bato is particularly insightful: "Koppel Fraenkel und seine Nachfahren. Die Schicksale einer deutsch-jüdischen Familie im Wandel von mehr als drei Jahrhunderten" ("Koppel Fraenkel and his descendants: The fate of a German-Jewish family over the course of more than three centuries"), *Israelitisches Wochenblatt* (Zurich), July 3, 1964.

[9]See Nathan Drori, Susan Edel et al., eds., *The Feuchtwanger Family: The Descendants of Seligmann and Fanny Feuchtwanger* (Tel Aviv: Feuchtwanger Family committee, 2009), printed by DoroTree Technologies (Jerusalem).**

[10]The J. L. Feuchtwanger bank, established in 1857, was liquidated with the rise of the Nazis, in 1937. In 1936, it was reestablished in Israel as the I. L. Feuchtwanger bank, which closed in 1967. Additional details can be found in Drori and Edel, ibid., pp. xxxii–xliii.**

In the *Yom Kippur* (Day of Atonement) volume, Abraham Fraenkel not only wrote his own name in German and Hebrew, but also the titles of the *Selichot*[11] selected for the *Shacharit*,[12] *Musaf*,[13] and *Mincha*[14] services, as well as the text of ויאתיו (*v'ye'etayu*), a *piyyut* (a liturgical song or poem). According to the mahzor, this was recited only in the *Musaf* for *Rosh Hashanah* (the Jewish New Year), but Cantor Abraham Fraenkel evidently also included it on *Yom Kippur*, as was the practice in Eastern Europe, and as far west as Berlin.

When Abraham Fraenkel was about 35, he married the young Nanette Neubauer. He must have already accomplished much in life, since his bride came from a respected and wealthy family and was well versed not only in Jewish knowledge, as was common for girls, but also in French and English. The marriage produced seven children, three of whom died at a very young age. The other four were still alive at the turn of the twentieth century: the firstborn Yitzhak Seckel Sigmund, born on August 16, 1827; Zechariah Benjamin Wolf (later Wilhelm), born on December 20, 1830, who was my grandfather; Koppel Jacob, born on October 9, 1833; and Rosl Rosalie Rosa, born on February 26, 1843, who was my grandmother and named after Abraham Fraenkel's mother. After the entry (in his *Chumash Derekh Selulah* mentioned above) for his firstborn, Abraham Fraenkel noted, וימל אברהם את יצחק בנו.[15]

My memories of my father's parents are all the more vivid, since, until my grandfather's death, which was most of my time in Munich, we lived together in the same building at 30 Klenzestrasse, close to Gärtnerplatz and the Gärtner Theater. The very old-fashioned house that belonged to my grandfather included two four-room apartments on each of the first two upper floors facing the street. The separating walls had been removed to convert them into two larger, but extremely uncomfortable, apartments for my grandparents and my parents. For two decades, with five children and attending servants, my mother suffered the ordeal of living in that apartment, with much hardship, but without complaint. The central building housed the offices of the "A. Fraenkel, Wholesale Wool Business," founded by Abraham Fraenkel. The warehouse for storing the bales of wool was across the courtyard. For us children, the most important space was the courtyard between the central building and the warehouse. It was spacious and made beautiful by chestnut trees. My siblings and I, and often our cousins too, played there in the summer, and built snow mountains and tunnels there in the winter.

---

[11]Repentance prayers and poems for *Yom Kippur* and the preceding weeks, as well as for other fast days.**

[12]*Shacharit* is the morning prayer service.**

[13]*Musaf* is the additional service, recited on the Sabbath, Rosh Chodesh (the first day of a new month), and on holidays.**

[14]*Mincha*: afternoon prayer service.**

[15]"And Abraham circumcised Yitzhak," Genesis 21:4, which refers to the entries in the first source mentioned on p. 2.**

When he married my grandmother Rahel Auerbach (1839–1915), my grandfather Wolf changed his name to Wilhelm Fraenkel (1830–1907). According to the Frankfurt authorities' marriage register, this took place on December 10, 1858. The marriage produced seven children, all born in Munich. Like many children at that time, their oldest son Adolf Abraham (1859–1868) died young of diphtheria. The other children were: Sigmund Aviezri[16] (1860–1925), who was my father; Heinrich (1862–1940); Toni (1865–1922), who married Abraham Auerbach of Cologne; Emil (1867–1942), the only one who was always in good spirits and was sensitive to all things poetic; Emma (1868–1928), who married Leo Mainz of Frankfurt am Main; and Berta (1875–1961), who married the physician Dr. August Feuchtwanger of Munich. The three siblings who still lived in Munich in 1933 emigrated to Haifa and Jerusalem between 1935 and 1939. Additional information about them, their children, and their grandchildren can be found in the aforementioned Auerbach family tree.

Rahel (third from left, seated) and Wilhelm (center, seated) Fraenkel with their children and children-in-law

Wilhelm and Rahel came from rather different backgrounds. Although his religious stance was Orthodox, it was far from the strict, militant stance Rahel brought from her parental home, which was absolutely unknown in Munich at that time, although Jacob Loew Feuchtwanger was not averse to it. They also differed

[16]The name Aviezri appears again among his descendants. It is the name of the grandfather (who died in 1767), and not the father, of his grandfather B. H. Auerbach. This can be explained by the fact that B. H. Auerbach's father was called Abraham (1763–1845), the name chosen by the paternal side for Sigmund's older brother, who died early [see the Fraenkel family tree in the appendix].

greatly in their temperaments. Wilhelm was the quiet, introverted type, best characterized by the attribute "wise." Thanks to his commercial foresight, he was able to develop his father's company from a local success into one of the most significant German trading companies in sheep wool. However, his interests were mainly intellectual. In his final years, his impaired vision made it difficult for him to read. Therefore, as a young high-school student, I used to read the daily newspapers to him in the evenings, especially the *Frankfurter Zeitung*, as well as Bismarck's *Gedanken und Erinnerungen* [Thoughts and Reminiscences].

This reading selection can probably be explained by two factors. First, the rather low standard of Bavaria's, and especially Munich's, printed media at that time, while the *Frankfurter Zeitung*, especially from an economic viewpoint, was the leading left-liberal, and later democratic, German daily newspaper, rather than the Berlin newspapers, *Vossische Zeitung* and *Berliner Tageblatt*. Second, even among left-liberals and "independent" citizens like my father, who viewed Reichstag representative Eugen Richter (1838–1906), one of Bismarck's leading critics, as their man, and considered the struggle for free trade, as opposed to protectionism, their main economic policy goal, the name Bismarck retained its glow. This was perhaps partly due to the profound conflict between Bismarck and Wilhelm II. In southern Germany, Wilhelm, the third and final German emperor, was viewed as representing a monarchy that was constitutional in name only. He also stood for reactionism and anti-Semitism, despite his contacts with Jewish figures, such as Albert Ballin and Gerson Bleichröder.[17] Indeed, he was considered the epitome of hated Prussianism, as compared to Bavarian particularism and southern German open-mindedness.

In contrast to her husband, a thoughtful man who closely followed all the political and economic developments, Rahel Fraenkel-Auerbach was a born fighter, both in the family circle and the Jewish community, especially championing her father's Orthodox tradition. Grandmother Rahel was a combination of stark opposites. She had a reputation for hospitality, and her care for the Jewish students from abroad was unmatched in the Munich of her day. She would appear early in the morning with a bottle of kosher milk at the lodgings of a student who was still in bed after having been out dancing all night and was used to committing more serious transgressions than enjoying non-kosher milk. Guests, particularly students, frequented their dinner table on Friday evenings, even though this was still rather uncommon in the Jewish Munich of my childhood. However, within the family, she was harsh and ungenerous. She was very strict with her children, and especially with the servants, with frequent turnover as a result. Not even the grandchildren were spared. Nevertheless, like my father, I enjoyed preferential treatment, since she probably assumed I would maintain the Orthodox family tradition. My grandfather loved me for very different reasons: my achievements in school and intellectual interests. However, "love" in those days, especially in my family, did not have any sentimental manifestations. I do not remember ever being kissed by parents or grandparents, and even words of praise were extremely rare. Not to be yelled at by my grandmother was the best you could hope for.

[17]Cf. Wilhelm II's pronounced acclaim for Friedrich Delitzsch's *Babel und Bibel*.

In connection with Rahel Fraenkel, I have to mention Dr. Heinrich (Chanoch) Ehrentreu (1854–1927).[18] Starting in the mid-nineteenth century, the Munich Jewish community was considerably strengthened by influx from the countryside, both from the Swabian communities of Krumbach-Hürben and Ichenhausen, as well as Franconian towns, such as Fürth. In addition, the community was strengthened by a general economic upturn. In 1887, the Jewish community finished building a magnificent synagogue near Karlsplatz, with an organ and a "mixed choir." At the same time, the community showed growing intolerance toward the "observant" (*gesetzestreu*) Orthodox minority. For this reason, the Orthodox minority formed an association, called officially the League for the Promotion of Jewish Scholarship (*Vereinzur Förderungjüdischer Wissenschaft*), but de facto Ohel Jakob (Jacob's Tent). In 1892, Ohel Jakob established a "private synagogue" with the same name on Kanalstrasse (later, Herzog-Rudolf-Strasse). By 1885, they had already appointed as their spiritual leader the "private scholar" Dr. Heinrich Ehrentreu from Hungary, who had served as the Bondi family's private tutor in Mainz until 1884. Although he was asked to succeed David Hoffmann as rector of the Berlin Rabbinical Seminary in 1922, Ehrentreu remained in Munich until his death. He was not only a first-rate Talmud scholar, but also a well-trained scientist who discovered and made use of extensive treasures in the Royal Court and State Library of Munich: manuscripts, incunabula, and old prints, including those of the uncensored Talmud. Due to the laws pertaining to Jews,[19] along with intolerance from the administration and the rabbi of the Jewish community at the time, Ehrentreu was denied the title of Rabbi and permission to give sermons, although he was a far superior Talmudist than the rabbi. His official role in the Jewish community was solely to supervise the ritual institutions. Finally, after bitter dispute, he was given the title of Rabbinical Assessor in 1907.

Rabbi Dr. Heinrich (Chanoch) Ehrentreu

[18]See the relationship to the Fraenkel family in the Fraenkel family tree.**

[19]Those laws permitted only a single unified congregation in each administrative district with only one rabbi. See discussion on pp. 24–26.**

Wilhelm Fraenkel's educational and business foresight, sharpened by extended business trips as far as Moscow, was made especially apparent by a highly unusual act at the time in southern Germany. He sent his sons Sigmund and Heinrich to Trieste and Marseilles for about two years to promote their personal and professional education and improve their foreign language proficiency before taking them into his business. Even more important for my father Sigmund was an extended apprenticeship he completed with the international company Aron Hirsch & Sohn in Halberstadt. Councilor of Commerce Benjamin Hirsch, the company's director, was married to Rahel Fraenkel's sister Julie.

While acquiring a general education, Wilhelm Fraenkel's sons also received Jewish instruction in Pentateuch and Talmud from a private tutor named Wolfsheimer. His daughters attended the "Institute," a secondary school for girls that focused on modern languages, and received intensive training in all areas of housekeeping from their strict mother. Oddly enough, their Orthodox mother also had her daughters trained in ballroom dancing, based on the general notion that young girls would not find husbands if they did not appear at the appropriate balls. Of course, this did not include the formal royal balls and masquerade balls, which were the focus of social life during the weeks of carnival celebration in Munich.

My grandparents' way of life was extremely simple relative to their financial status. Only on the evenings of Simchat Torah and Purim festivals were all the children invited to attend festive and abundant dinners. These dinners always included the punctual arrival of a turkey sent by sister Julie from Halberstadt, who, in turn, received from us specialties from Munich, such as lemon pears, russet apples, and Merano grapes. Whereas sea fish were not available in Munich until the turn of the twentieth century, and were not all that eagerly welcomed by the Bavarian palate, there was no lack of delicious fish from streams and foothill lakes. Now and again, my grandparents took a spring vacation away from the harsh Munich climate, with snowfall sometimes starting even before *Yom Kippur* and continuing until May. They would vacation in the warmer Merano region of northern Italy, and even travel as far as the Italian Riviera. My grandfather told stories to the grandchildren, who would sit on his knees and listen, all ears, about how the first "foreign" sausages were sold at the train station in Kufstein and how the "southern" climate started somewhere between Franzensfeste (Fortezza) and Bolzano.

Rahel's religious influence on her husband and children was substantial. First of all, of course, it affected their lifestyle both inside and outside their home, including the employment of Jewish cooks, most of whom were from Franconia. These cooks and several Christian maids and nannies helped out, but Rahel never relinquished charge of her kitchen or household. Her influence was also evident externally. For instance, her husband was elected to the Jewish religious community's administrative council in 1875, where he remained until 1902, when my father took his place. He embraced all the observant Orthodox minority's interests, energetically and wisely. Otherwise, he led a withdrawn life, focusing on family, his business, and reading, leaving the social activities to his wife.

For most of the time, my maternal grandparents lived in Ruffini House, an old and spacious building not far from Marienplatz, until it was torn down around the turn of the twentieth century.

My mother's father, Isidor (Israel Arjeh) Neuburger, was born in 1837. A wealthy young man, he moved from Fürth to Munich and joined the Fraenkel wool business. He soon became a partner, and married Wilhelm's sister Rosa. Owing to his considerable share in the business, he received a very sizeable portion of the company's profits, and, toward the end of his life, was considered a millionaire. In the 1890s, he left the business, partly due to disputes with his brother-in-law Wilhelm, and lived as a "man of independent means." My memories of him are superficial. He led a strictly Orthodox life, but his attitudes were more like those of most Jews who came to Munich from the Bavarian countryside, and he had few intellectual interests. After the turn of the century, he became sick, and died in the summer of 1905, my only grandparent who did not reach the age of 70.

His wife Rosa (1843–1927) was the absolute antithesis. Far more intellectual than her husband, she was like her brother Wilhelm in many ways. As far as I can remember, aside from Flora Sassoon and Shoshana Persitz, née Zlatopolski, my relationship with her was the most significant and closest with a woman in the course of my life. Her marriage in 1864 was a marriage of convention, based on business considerations, as was common at the time. Rosa was deeply religious, but, although devoted to the traditions of her parental home, she avoided the militant Orthodoxy of her sister-in-law Rahel. Her occasional concern over possible damage to our relationship because of this proved to be utterly unwarranted. Eminent men sought her company, including the laryngologist and cardiologist at the University of Munich, Prof. Max Oertel, who developed "terrain cures," and, later, the bibliophile Maurice Ettinghausen of Rosenthal Antiquarian Booksellers. Ettinghausen was an English Jew who traveled widely in the Far East, and whose son Walter (later, Walter Eytan) became the Israeli ambassador to France.

Rosa always knew how to give good and wise advice, and close and extended family would turn to "Tante Rosa," even though her harsh and uncompromising manner made her difficult to get along with. In over two decades of widowhood, she by no means relied exclusively on her children and grandchildren. Until the First World War, she was a wealthy woman. Since, as a young woman, she had run her brothers' household in Vienna, she was an incomparable cook, and was thus both an intellectual and a culinary attraction. In 1925, after my father's death, when she was over 80 years old and ailing, she moved in with my mother for the two remaining years of her life, much loved by her grandchildren and great-grandchildren.

Two of Isidor and Rosa Neuburger's five children succumbed rather young to diphtheria and measles. The other three were: my mother Charlotte (born in 1868); Albert Neuburger (1878–1958), who had twins from his marriage to Jeanne Offenbacher; and Therese (1884–1940), who married the distinguished mathematician Alfred Loewy (1873–1935) in 1906.

The allegiance of my mother's siblings to the German people and country ended tragically. With unshakable confidence in a German victory in the First World War, they lost their considerable wealth by investing in German government securities.

However, the real tragedy began in 1933. In the spring of 1934, when I accompanied my mother to Jerusalem, I met up with her in Switzerland, not daring to travel to Germany. Her brother and sister were opposed to breaking up the family and abandoning the "fatherland." Her brother, having survived the ordeal of Dachau concentration camp in late 1938, then made his way to Palestine/Land of Israel.[20] However, her sister, who had lost her husband in early 1935, after his dismissal from the university soon after the Nazis took power, refused to accept the emigration certificate for Palestine we offered her. In her letters until 1939, which in the face of reality were rather naïve, she described how comfortable she—an observant Jew—still felt in the circle of her "Aryan friends," some of whom took instruction from her in singing and breathing techniques. But disillusionment came quickly, and, in the fall of 1940, when the Jews of Baden were forcibly deported to the Pyrenees, she slit her wrists. She was the only victim of the Nazis among my close family. Except for the few who fled to England or the United States, they were almost all, especially all my siblings and their families, reunited in Palestine when the war began.

Sigmund Fraenkel

It is not easy to describe my father Sigmund adequately, since he was not only talented and notable, but also an unusually multifaceted man. Some idea of his scope is conveyed by his selected publications and speeches, which I published in 1930, in Jerusalem, under the title *Sigmund Fraenkel. Aufsätze und Reden. Ein Spiegelbild deutsch-jüdischer Geschichte aus dem Beginn des zwangzigsten Jahrhunderts* (Essays and Speeches: A Mirror of German-Jewish History from the Early Twentieth Century).[21]

---

[20]During the Ottoman Empire and the British Mandate, the area that later became the State of Israel was known as Palestine or Palestine/Land of Israel. For the sake of clarity, this book will use the formulation "Palestine/Land of Israel" to denote the region, and *yishuv* (Hebrew "settlement") to refer to the Jewish settlement or community there from 1860 to 1948, prior to the founding of the State of Israel.**

[21]A. A. Fraenkel, *Sigmund Fraenkel. Aufsätze und Reden* (Munich: B. Heller, 1930).

Sigmund Fraenkel spent six years at the Ludwig Gymnasium, a humanities-based academic secondary school in Munich, and was known throughout his life as a good Latin scholar. He completed his apprenticeship in Halberstadt (1877–1878), and, later in Trieste and Marseilles. Working for Aron Hirsch & Sohn, a company with an extensive network of international contacts in the United States, Australia, and the Far East, enhanced his business skills. At the same time, he was greatly impressed by how the company combined global business with Jewish tradition. On *Yom Kippur* eve the company's chief executive, Uncle Benjamin Hirsch, would ask all his colleagues and employees to forgive him for any grievances he might have provoked during the year. Another example arose when the First World War started on August 1, 1914, on the Sabbath, as is generally known, which was also the eve of Tisha b'Av, the day of fasting to commemorate the destruction of the temples in Jerusalem. Although Aron Hirsch of Berlin, the firm's general director of the industrial companies (Messingwerk brass foundry, Ilsenburg copperworks, etc.), did not observe traditional Jewish practice, he was very interested in Judaism. He called the company in Halberstadt after the Sabbath ended to discuss important war-related decisions with Dr. Emil Hirsch, the chief executive of the trading company at the time. Told that Emil was at the synagogue to hear *Eicha* (the prophet Jeremiah's Lamentations), he responded: "Tell Dr. Emil that people in Berlin have already been reciting *Eicha* since Saturday morning!"[22]

My father's intellectual development during his stay in Halberstadt was greatly influenced by his uncle, Dr. Selig Aviezri Auerbach (1840–1901). In 1872, Selig succeeded his father as the rabbi of Halberstadt and director of the Jewish school and a small yeshiva (Talmudic college). Uncle Selig, whom I still vaguely remember, was an unusual personality. As the congregational rabbi and Talmud teacher to whom students came, even from Berlin, he combined great talent as an educator and Orthodox strictness with an openness that was lacking among most of the next generation. He earned his rabbinical ordination from Rav Yitzchak Elchanan Spektor of Kovno and was head of the board of Azriel Hildesheimer's rabbinical seminary in Berlin, which was opposed by the Frankfurt Orthodoxy. For a while, he was also the president of the General Rabbinical Association of Germany (made up of both Liberal and Orthodox rabbis). In protest of the religious neutrality of the General Rabbinical Association of Germany, both "Orthodox" and other "observant" Jews established separate rabbinical associations.

His uncle's spirited Gemara study and the Orthodox style of Jewish life in Halberstadt, which lacked even a hint of fanaticism, left a profound, lifelong impression on my father. A few characteristic traits illustrate his kind of religiosity. When my brother and I accompanied my father to the Kol Nidre service in the years after Bar Mitzvah, he would say: "The inner harmony with our Creator with which we recite our prayers tonight does not exist among other peoples and religions." As he impressed upon us, if any of his children were to abandon the Jewish tradition, it

---

[22]He meant that they started mourning at the coming of war on the previous day. Note that this episode took place many years after the apprenticeship of Sigmund Fraenkel.**

would give him the greatest pain and ruin his life. Even on his deathbed, his religious conviction, though mixed with self-reproach, outshone all his other feelings. However, he told my brother and me that we could freely choose our professions, except for becoming rabbis, reflecting his disappointment with many rabbis in Bavaria and northern Germany. He also did not like us to cover our heads with skullcaps or hats except for prayers and at meal times, which was in keeping with the halacha, but contrary to the widely popular custom. Furthermore, he disliked divergence from the non-Jewish environment in any way that seemed unnecessary to him.

But the decisive, lifelong, and most significant impact of the example set in Halberstadt was Sigmund Fraenkel's insistence that all his sons receive intensive Talmudic training, starting from their ninth birthdays. This also greatly influenced his daughters' sense of "learning." Even before the turn of the century, the Orthodox lifestyle began to spread in Munich, starting with the Fraenkel and Feuchtwanger families. In the Feuchtwanger family, this was promoted particularly by Lothar Feuchtwanger (1856–1913), who was Jacob Loew Feuchtwanger's son, and Dr. Elias Straus's uncle. This new form of Jewish life was later strengthened when other families moved there, especially from Franconia.

Selig Auerbach married Röschen (Little Rose), Benjamin Hirsch's sister (1843–1932), who survived all the profound twists of fate that struck the Hirsch family until shortly before the Nazis took power. We all adored Tante Röschen. She had 12 children, and 232 living descendants in 1956, as compared to 96 from Rahel Fraenkel. In both cases, most of them lived in Israel.

Julie (Julchen) Auerbach (1844–1916), my grandmother Rahel's youngest sister, married Benjamin Hirsch (1840–1911) in 1864. He was the son of the Councilor of Commerce Josef Hirsch, and grandson of the Aron Hirsch & Sohn company's founder. A brilliant businessman, in 1906, together with the younger Aron Hirsch, he founded the company's industrial branch: Hirsch Kupfer- und Messingwerke (Copper and Brass Foundry, HKM) company, in Eberswalde and Ilsenburg. Under his guidance, the Halberstadt company developed into an international firm with subsidiaries around the world. He also became a Prussian Councilor of Commerce and the first non-British member of the London Metal Exchange. As for training, Uncle Benjamin, whom I remember vividly, was essentially a self-made man. He does not seem to have had any formal schooling, aside from studying for two years at the Eisenstadt yeshiva, started by Rabbi Azriel Hildesheimer, also from Halberstadt, who later founded the Berlin Rabbinical Seminary. However, he acquired Jewish and secular knowledge by other means. His two older sons studied chemistry, receiving their doctorates before joining the company. His youngest son Joseph, who was a gynecologist in Berlin, moved to Jerusalem in 1933. Benjamin Hirsch strongly influenced my father, with whom he traveled to New York, in 1905. The following story is characteristic. The captain of one of the fast steamers of the North German Lloyd (NDL) shipping company, with which he often traveled to the United States, invited him to the command bridge to chat. Benjamin Hirsch pulled out his *luach* (Hebrew calendar) from his pocket and said, "Tonight at such-and-such an hour and minutes is the new moon." The captain answered, "If we were to navigate according to your new moon, Mr. Councilor of Commerce, the ship, along with all hands, would be lying at the bottom of the ocean." Benjamin did not know that the new moon times (*moladot*) listed in the *luach* were only the mean values, which could deviate from the true times by many hours, or even a full day. Unfortunately, this fact is generally not taken into account in synagogues even today.

When my grandfather Wilhelm Fraenkel died in late fall 1907, Uncle Benjamin and Aunt Julie came to Munich for several days, despite the crisis that had just erupted on Wall Street. I would like to mention that at the turn of the century, the three leading German non-ferrous metal (copper, brass, etc.) companies were all in Jewish hands and founded by Jews: Aron Hirsch & Sohn in Halberstadt; Beer, Sondheimer& Co., run at that time by the chief executive Councilor of Commerce Louis Feist, in Frankfurt am Main; and Metallgesellschaft, run by Councilors of Commerce Leo Ellinger and Zacharias Hochschild, also in Frankfurt am Main. The first two companies were run by strictly Orthodox Jews and closed on the Sabbath.

After his general business education in Halberstadt, my father spent a training period abroad in the special field of wool, mostly in Trieste, which was then part of Austria. I have many letters my father wrote to his parents from both these locations. Today, the industriousness, Spartan simplicity, and sacrifice for the sake of kashrut, which characterized his life during those years, seem incredible. Leave to visit his mother was the only reward he desired. Almost all his letters express his touching devotion to his parents and siblings, and his efforts to take advantage of the business training while minimizing expenses.

Although he would have liked to explore other countries and further his education, he considered it his duty to enter his father's business when he was 20 years old. My father gained specialized expertise in southern German wools, and was also an autodidact, broadening his knowledge in economics, Jewish history, and literature. Particularly interested in geography and commerce, he spent decades as an active member of the Munich Geographical Society, and became friends with Dr. Siegmund Günther (1848–1923), a geography professor at the Technical University of Munich since 1886. This relationship also had an important impact on my own development and remained untarnished until the end of the First World War, when the "stab-in-the-back myth"[23] and family influences drove Günther into the anti-Semitic camp. It was typical of my father's attitudes in the early twentieth century, when I was 12 or 13, that he imagined that I would become a geography professor. At that time others in our circle warned their sons against pursuing academic careers, which seemed rather hopeless for Jews. I remember being afraid to ask him to suggest a possible topic for a doctoral thesis in geography, about which I only had a very rudimentary high-school knowledge. Encouraged by my

[23]The stab-in-the-back myth is the anti-Semitic and anti-Republican slander that appeared in Germany after 1918, following Germany's defeat in the First World War, claiming that the German Army did not lose the war but rather was betrayed, or stabbed in the back, on the home front, especially by Jews.**

father to think about this on my own, after a while (60 years ago!), I came up with the idea of irrigation projects for the Sahara Desert.

My father was even more intensively preoccupied with questions of commerce and transport. One of my father's speeches was "The Bavarian State Railway Network and the Overland Route,"[24] which he gave to leading economists and politicians at the Commercial Association (KV) in 1891. With precise calculation, and references to distances and train and ship speeds and tariffs, he proposed channeling the traffic in the first stage of the London–Port Said–Aden–Bombay route via Ostende–Vienna–Saloniki, rather than via Calais–Mont Cenis–Brindisi, the route until then. In addition to traffic advantages, it would yield considerable income for the Bavarian Railway Administration along the Aschaffenburg–Passau rail route. His talk ended with the slogan: A good patriot will above all be a good railway man.

My father retained an interest in commerce and transport throughout his life and was appointed a member of the Bavarian and later the Reich Railway Council. He passed this interest on to his sons. By the time I was 11, I had already memorized the schedules of the express trains on the main railroad routes in Germany and Austria, and worked out travel plans to various health resorts for our friends. Until 1929, when I left Europe, I immersed myself in the annual summer schedules of the central and western European express trains much more than the year's new literary releases. The same went for my brother.

My parents knew each other as cousins and lived in proximity to each other from earliest childhood. They married in November 1889 when my mother was 21. The marriage produced five children: I was the eldest (born on February 17, 1891), Eugen Joël (in March 1892), Else Lea (in July 1896), Paula Perl Tirza (in May 1901), and Dora Dvora (in November 1904). Else, a mother of four, died at the age of 30 from an infection she presumably caught from a child with scarlet fever. All the others are married and live in Jerusalem. Paula has eight children.

At the time of his marriage, my father was a partner in the A. Fraenkel Company. His senior partners were his father and his father-in-law. Subsequently, his brother Heinrich also became a partner, but suffered his whole life from the constraints of business life. He would have preferred to pursue an intellectual profession.

My father was active in the first B'nai Brith lodge in Bavaria from its founding in 1897 in Munich. It had 127 brothers (members), which was quite a respectable number for that location. My father especially supported starting a Jewish "nurses' home" (*Schwesternheim*, 1900), which soon grew into the Munich Jewish Hospital ("home for the sick"). Reflecting the high esteem he enjoyed as an active brother, and more so as a brilliant speaker, he gave the keynote addresses at the 7th, 10th, and 25th lodge anniversary celebrations. This is all the more remarkable in view of the sharp clash of religious factions in the Munich Jewish community, with only a small minority of lodge brothers being Orthodox Jews, including my three Fraenkel and Feuchtwanger uncles. Many Ohel Jakob members, including some

---

[24] This lecture is included in his collected works mentioned previously (see note 21).**

Feuchtwangers, avoided the lodge for religious reasons, so as not to have to work together with the religious Liberals.

Sigmund Fraenkel emphasized the lodge's significance not only in strengthening Jewish self-awareness and combating anti-Semitism, but also in showing solidarity with the *Ostjuden* (Eastern European Jews) in Munich, who were held in rather low esteem by most long-established Munich Jews. While this obligation was clear from the outset in the American branch of B'nai Brith, the "Liberal" Munich brothers did not like to hear about it.

My father's lodge speeches have almost all been published. They are notable for referring to both secular and religious sources, which today seems somewhat strange, although it was typical for broad circles of German Jewry back then. This was particularly pronounced in his speeches. On the one hand, verses and entire works by German poets (Goethe, Schiller, and others), as well as Richard Wagner, were used as leitmotifs. On the other hand, they were supplemented by biblical, halachic, and especially aggadic quotations. For example, the speech for the tenth anniversary has references to the second act of Tannhäuser and also to the *Midrash* about the three gifts (the well, pillar of cloud, and manna) the people of Israel received during their desert wanderings, through the merits of the siblings Miriam, Aaron, and Moses.

Particularly noteworthy is the 25th anniversary speech, because it is based entirely on Richard Beer-Hofmann's poignant drama *Jaákobs Traum* (Jacob's Dream), which is shaped by Jewish national-religious motifs. My father was friendly with the Viennese poet-dramatist who wrote the play. In 1921, the year before he made the speech, he had seen the play, together with the lodge's president, at the Deutsches Theater in Berlin.

The Munich lodge founded several significant institutions, amongst them Toynbee Hall, a center for educational and social services, in addition to the nurses' home and hospital, mentioned previously. However, my father turned down the opportunity to stand for election as president of the lodge, only partly because he had so many other obligations. Another lodge experience weighed heavily on him. My father had recommended Dr. Elias Straus (1878–1933),[25] the oldest son of Samuel Straus of Karlsruhe, as a new brother for the Munich Lodge. Straus was a highly talented lawyer and an ardent and knowledgeable Jew, albeit not strictly Orthodox. He was one of the earliest active Zionists in Bavaria and was acquainted with Chaim Weizmann. According to the secret ballot system, which was, and still is, used in the European lodges and in Israel, but not in the United States, home of the B'nai B'rith, only a few black balls were needed to exclude a candidate. Despite my father's warm recommendation, Straus was blackballed from joining the lodge because of his Zionism. Indeed, after Munich's Jewish community thwarted Theodor Herzl's plans to hold the first Zionist congress there, the city gained a reputation as the stronghold of anti-Zionism in southern Germany.

---

[25]See also the book by his widow, Rahel Straus, *Wirlebten in Deutschland*, published in association with the Leo Baeck Institute (Stuttgart: Deutsche Verlags-Anstalt, 1961).

Since I showed no interest in the lodge in Germany, neither my father nor I would have ever dreamed that in 1930 I would join the Jerusalem lodge as a brother. Moreover, after the State of Israel was founded, I was elected the first president (1949–1952) of the Israel district (previously Palestine district) of the B'nai Brith lodge.

In a long letter I have, dated August 29, 1900, from Scheveningen, my father wrote, as always by hand,[26] to my mother: "Dear wife! You have surely been very angry with me all day because people have elected me into an office for which they consider me to be particularly qualified, and to which, I will not deny, I did aspire a little, in spite of everything. But I promise I will delegate a lot of things I have so far needlessly taken upon my shoulders. So you should also be a bit proud to be the wife of a not-yet-40-year-old member of Bavaria's first trade association. At least I am pleased I cannot hear or see your anger over the cards congratulating me that will come in the mail. I know that at the bottom of your heart you are also a little pleased, if only for the sake of our beloved children. I would also like to believe that, while your beloved Papa will respond with a grumble to the congratulations that come his way today in the stock exchange café, he will feel a sense of inner satisfaction, nonetheless."

My father's election to the Upper Bavarian chamber of commerce signified a new focal point in his life, in addition to his business- and Jewish-related (thus always also literary and political) activities. His very active family life was a major focus which formed the basis of his life. The work in the chamber of commerce led to diverse activities, which continued for the next 25 years, until his early death. He presented many reports from 1900 to 1924 on fundamental questions of Bavarian and German economic policies, which were published in magazines and newspapers. Regarding commercial issues, he represented the interests not only of the merchant community, but also of the Bavarian, and later the Reich, governments. Many of his proposals and essays in that area appeared in the *Frankfurter Zeitung*. His untiring labors and the success of his efforts were recognized in 1909, when he was elected vice president of the chamber of commerce, and, soon afterwards, conferred the title Councilor of Commerce. To be elected vice president of the chamber of commerce was extraordinary for a Jew, but the position had some constraints. According to an unwritten law, the presidency of the chamber of commerce was reserved for a Catholic representing a typically Bavarian branch of business. In 1924, the Minister of Trade recommended my father for the rare title of Privy Councilor of Commerce (*Geheimer Kommerzienrat*), in keeping with Bavaria's obsession for titles, which it retained, in contrast to northern Germany, even after the German Revolution of 1918.[27] However, Minister-President Heinrich

[26]Even after the A. Fraenkel company office acquired a typewriter, in the early twentieth century, Sigmund Fraenkel continued to write all his letters to his family and many others by hand.

[27]In the wake of the First World War, in 1918, the November Revolution in Munich led by Kurt Eisner aimed to establish a socialist Free State of Bavaria. Bavaria's King Ludwig III was forced to abdicate, followed by the rulers of the other German states as the revolution spread and republics

Held, head of the government of Bavaria from 1924–1933, declared to the Council of Ministers that such an appointment would be too great a provocation for the National Socialists.

It was typical of Sigmund Fraenkel that he never shied away from introducing Jewish-religious ideas into his economic policy considerations, even using them as essential arguments, although this sometimes required considerable courage. In his comprehensive treatment of "the future war tax on property,"[28] for example, he wrote: "When I think about such problems, I take refuge in a reference book with the advantage of also being the oldest textbook in human history. I take refuge in the Bible." Then he analyzed the social rules of Leviticus and Deuteronomy referring to tithes, and basing his comprehensive emergency tax program on them.

Also, in the 46-page series of articles published in 1925–1926 on "the seven Egyptian years of famine"[29] (in the book of Genesis), he incorporated modern economic ideas as an aid to a detailed understanding of the policies of Egypt's "economic dictator," Joseph. He also included extensive discussion of medieval Jewish exegeses.

He was able to take advantage of his influence in the Chamber of Commerce to support Jewish interests. For example, he successfully supported authorization for trade companies observing the Sabbath to do a limited amount of office work on Sunday.[30]

Albeit not explicitly based on Jewish ideas, his open attitude, maintained even under difficult circumstances, favored cooperation with Western Europe in general, and Great Britain in particular. For example, when the German Emperor Wilhelm II fueled the conflict with England with the Krüger telegram in the Boer War (1896–1902), the official stance of the Germans, including the Jews, was definitely on the side of the Boers and against England. However, I vividly remember my father convincing me of the opposite view when I was still very young. He claimed that England was the most liberal country, where persecuted Jews always found asylum. Likewise, at the height of German-English economic disputes, he enthusiastically advocated reestablishing the "old friendly relations between the two great cultural nations."[31] Some time earlier, he explained to us the essence of the Dreyfus affair, Emile Zola's *J'accuse* of 1898, and the reasons why a pardon was not an adequate solution.

His letter written in late 1917 to the weekly *Handel und Industrie* about Germany's aims in the First World War shows his characteristic courage in defending his political views. Only from material found by the Western Allies in

---

were declared. A National Assembly in January 1919 later led to the establishment of the Weimar Republic in August 1919. See also the beginning of Chap. 5.**

[28]"Die zukünftige Kriegssteuer des Besitzes," *EuropäischeStaats- und Wirtschaftszeitung* (1916), no. 6.

[29]"Die sieben ägyptischen Hungerjahre," *Jeschurun* 12 (Berlin, 1925), nos. 5–10.

[30]See *Bayerische Handelszeitung* (1912), no. 16.

[31]See *Bayerische Handelszeitung* (1906), no. 2.

German archives after the Second World War did the extent to which Wilhelm II and Erich Ludendorff planned to enslave the world after their envisioned victory become known.[32] However, particularly in the winter of 1917–1918, and after the major German offensive at the French front around Amiens, in March 1918, the German centrist and right-wing parties again made public "German victors'" claims, in *vaevictis* ("woe to the vanquished") style. At that time, it was dangerous to swim against the political and military tides. Nevertheless, in his letter, Sigmund Fraenkel wrote: "[The remarks that] reconciliation and understanding between us and the Western powers is impossible in the future; that 'we cannot rest until the final strike toward the total submission of our western enemies has been completed;' and that 'there can be no other peace with England and France than in a church graveyard,' [are]. . .diametrically opposed to my conviction that an end to the present bloodbath. . . depends on the conclusion of a peace with understanding and reconciliation between us, the Western powers, and America. . . . I declare my support of the view that our economic life can only be rebuilt if these peoples start once again to trust one another. . . . I am well aware that my opinion is not currently shared by broad circles of industry."

With respect to this letter, my father—in contrast to my brother and me—was brought up in a Germany that was a rapidly rising power, and, in the early years of the war, he believed in German victory. Also, he was among the small minority who did not lose trust until unrestricted submarine warfare was resumed under the pressure of Admiral Alfred von Tirpitz. Sigmund Fraenkel anticipated that this would lead the United States to enter the war, based on the United States' power and youthful élan he had experienced with his own eyes while visiting there.

Only a single aspect of Sigmund Fraenkel's successful business activities is significant for posterity, his grandchildren, and great-grandchildren: his business trips abroad. Those in central, western, and northern Europe are not worthy of any special attention. He traveled to London for the wool auctions twice a year, and also regularly to Budapest. His trips to America, in May and June of 1905, mentioned previously, included a four-week stay in New York, Boston (the center of the wool trade), and other cities in the northeastern United States. On the ship voyage going there, he wrote a 64 quarto-page letter to his family. He also vividly described his later travel impressions.

However, four or five longer trips to Russia, between 1890 and 1913, for which he learned the rudiments of the Russian language, are of particular interest, especially with regard to the Russian wool market, with its main centers in Moscow, Kiev, Tsaritsyn (later Stalingrad), Odessa, and Rostov-on-Don. Of course, his letters and stories were mostly about Jewish life and communities in Russia. For example, he wrote about the two major synagogues in Odessa, which left a great impression on him. His stories were largely about the country's special features that were surprising to western European Jews. Although he enjoyed preferential

[32]See Fritz Fischer, *Griff nach der Weltmacht* (Düsseldorf: Droste, 1961) [English edition: *Germany's Aims in the First World War*, trans. C.A. Macartney (New York: Norton, 1967)**].

treatment on special recommendation from the German foreign office as a "first-guild" merchant, it was an arduous task for him to arrange his visa and hotel bookings. In particular, he did not want to "bribe" the authorities, since he was entitled to proper treatment in any case. In this he was, like his descendants, a true *Yekke*. As younger people may not realize, until 1914, travel within Europe and even beyond did not require passports or visas, with the exceptions of Russia and Romania. In those two countries, both native and foreign Jews were subject to special legal treatment. Passports stating whether the bearer was Christian or Jewish were required to obtain entry visas. Those issued to Jews were only valid in certain areas and cities, such as in the so-called Pale of Settlement. Also, police authorization was required to leave a city. There were no such problems outside those two countries. For example, on a sunny morning in Rome, someone deciding to take a trip to London or Stockholm just had to go to the train station and buy the appropriate ticket. The numerous national borders went almost unnoticed, except in some cases for customs inspection of luggage and a change in language.

My father's experience in Russia was typical. Both native and foreign Jews were only allowed to enter the city of Rostov-on-Don with special permission. However, when he went to the kosher restaurant there on Friday evening, he met more than a dozen Russian Jews who were staying in the city "illegally." After they finished their soup, the doors opened and a police officer in uniform came in. My father was startled, imagining they would all be arrested, except for him. However, the police officer calmly sat down at the table to eat a typical Jewish Sabbath feast of gefilte fish and cholent. As it turned out, he received a monthly salary from the restaurant owner, which included the Friday evening meal.

Our mother and we children were particularly unsettled by the following political developments. Sigmund Fraenkel was outraged over the willingness of the European governments to offer information about the religion of passport bearers applying for Russian visas at the Russian embassies, which would have discriminatory consequences. Thanks to his connections, my father got the foreign office in Berlin to refuse to provide such information, so that the Russians had to figure out on their own whether applicants were Jewish or not. This change in a long-standing diplomatic custom caused quite a stir, and, as a result, the French Chamber of Deputies introduced the following interpellation: If the German government, whose relations to Russia were never especially cordial, refuses to give the Russian embassy in Berlin information about its citizens' religion, surely, the French government, which had close relations with the Russian government, can follow suit. Fortunately, our fears that the Russian police authorities might retaliate against my father on his next trip for initiating this were groundless. At the beginning of the twentieth century, the United States government demanded that Russia treat all American citizens equally, irrespective of their religion; when Russia rejected this, the United States terminated the US–Russian trade agreement.

It would go beyond the scope of this book to present all my father's activities for Jewish causes, so I have limited myself to the main ones. Among the speeches he gave in the Munich lodge was an extensive eulogy for Rabbi Moses Cossmann Werner. This is notable because, as a leader in the Munich Orthodox community, he was honoring the memory of the Liberal rabbi of the unified congregation, eloquently expressing the sympathy and tolerance between the two men.

Before 1914, Sigmund Fraenkel gave several speeches on Jewish scholarship to the Munich Association for Jewish History and Literature, on subjects such as the English ghetto novella (especially Israel Zangwill), the destruction of Jerusalem in *midrashic*[33] poetry, the *selichot*, and the Vilna Gaon (Elijah of Vilna). His talk on the *selichot*, although based on Leopold Zunz and others, reveals an instinctive empathy for all things poetic. This confirmed our family's common impression that, behind my father's harsh, intellectual, strong-willed, often aggressive, and ironic exterior, lay a soft, emotional, sometimes almost sentimental, and always profoundly religious core. He passed on to me his particular love for the *selichot*, which are an essential component of the medieval poetry in synagogue service. In my own speeches and articles, I later often expressed my regret over the loss, especially in Eastern Europe and Israel, of the enchanting rites of Ashkenazi Jews linked with the *selichot* on the Day of Atonement, due to misunderstandings and indifference in the late eighteenth century. Although that ancient tradition was maintained in southern and western Germany, as well as in Alsace and Holland, those Jewish communities were almost entirely wiped out during the Nazi dictatorship.

Even more characteristic was the talk about Rabbi Elijah, the Vilna Gaon (1720–1797). As incredible as it sounds, before 1914, there were hardly a dozen "native" Jews in Munich who knew more about him than just his name. This was because, while Hungarian Talmudic scholarship made its way to Germany by 1920, Lithuanian and Russian-Polish Talmud study did not. The few exceptions, such as Salman Baruch Rabinkow, the eminent Talmud scholar and founder of the so-called Heidelberg School, simply prove the rule. Thus, Sigmund Fraenkel could not delve too deeply into the actual material, for example, by comparing the Lithuanian *derekh limmud* (method of learning the Talmud) with the Hungarian counterpart, which emphasizes the *Acharonim*. By presenting the Vilna Gaon not only as a thinker and researcher, but also, more importantly, as a person, he familiarized his audience with this unique personality who had initially been considered foreign and unknown.

This is perhaps the appropriate place to mention something about the *Ostjuden* in Munich. These Jews came from Galicia and Congress Poland, as well as some from Russia, and settled in Munich between the turn of the century and the First World War. Their numbers hardly ever exceeded a thousand. Not only the non-Jewish Bavarians but also the local Jews considered them incomparably more foreign than their counterparts in Berlin. In part, this was because of the small number of local Jews, which never noticeably exceeded 10,000. In addition, as a result of declining migration from Old Bavaria, south of the Danube, and the lower birth rates, the community's size remained almost unchanged throughout the first quarter of the century. The total number of Jews in Munich was roughly 8700 in 1900, and 10,000

[33] *Midrash* refers to halachic or aggadic explanations relating to the Bible.**

in 1925, among a total population of 500,000–600,000. In contrast, Berlin had 108,000 Jews in 1900 and 173,000 in 1925 (that is, 4.3 percent of the total Berlin population; and more than 5 percent in 1890), accounting for more than 30 percent of all German Jews. Moreover, Berlin was far more open to the "East" than Munich was, especially since Bavaria and Austria, but not Galicia, of course, were often regarded as a unit by Jews and non-Jews alike. Thus, the large majority of Munich Jews showed not only apathy, but often also animosity, toward the Jews from Eastern Europe, which is understandable, though not excusable.

Dr. Elias Straus and my father were among the few exceptions. In the southeastern part of the city, where many *Ostjuden* lived, a *Ritterstiftung*, a soup kitchen, was founded around the turn of the century. I have vivid memories of my father showing up there year after year on the evening of the first Passover *Seder*, on his way home from the synagogue. He would greet the Eastern European Jews in a way that drew attention to the significance of this night and to the sense of belonging and mutual responsibility among all Jews.

To understand Sigmund Fraenkel's Jewish political activities, it is necessary to outline the legal and actual situation of the Jews in Bavaria in general, and in Munich in particular. Until the First World War, it was determined by the Bavarian Edict of 1813 pertaining to the Jews. At that time, despite Bavaria's membership in the Napoleonic Confederation of the Rhine, full emancipation of the Jews had not been implemented, and there was not yet a religious, dogmatic schism among the Jews (in Bavaria!), nor was it even anticipated. The Bavarian state government repeatedly planned to revise the "Jewish Edict," especially in 1831 and 1848. The plan for a "Jewish Church Council for the Kingdom of Bavaria," based on the Württemberg and Baden models, was abandoned since, by about the middle of the century, the division of Judaism into Neolog and Orthodox camps had progressed so far that placing them both under a single authority with a synodal council structure was no longer possible. Furthermore, the ultramontane "Center" party, which was dominant in the parliament and government, was solely interested in maintaining the status quo, so as not to evoke suspicions of "Church reform." The following example serves to show the ridiculous consequences of this. When I entered academic secondary school (Gymnasium) in 1900, I already had some knowledge of Hebrew and also the Talmud. At the time, my father attempted to get me exempted from state-run Jewish religious instruction, which was given by a rather ignorant Neolog teacher from the Jewish community, who could not offer me anything I did not already know. His efforts were in vain, since the Ministry of Education and Cultural Affairs was fundamentally opposed to any "dispensation from religious instruction," which was among the six subjects with a written examination for the *Abitur* (university entrance qualification examination).

According to the central and most controversial point of the old "Jewish edict," it was compulsory for all Jews in a particular district to belong to the congregation. Therefore, leaving the congregation was tantamount to abandoning Judaism. Both the Neolog and Orthodox minorities in Würzburg, Munich, and Nuremberg considered this a coercion of conscience. In Prussia, the struggle against what was regarded by the Orthodox as coercion led to the Secession Bill of July 28, 1876. Introduced by Eduard Lasker (1829–1884), a national-liberal Jewish parliamentarian in the Reichstag, it gave religious minorities the option to secede from a congregation "because of religious reservations" and form their own secession congregations. The historically most significant applications of this law with the greatest impact were the "Israelite Religious Society" (IRG), the secessionist Orthodox congregation in Frankfurt am Main, under Rabbis Samson Raphael Hirsch and Salomon Breuer[34]; and the Adass Yisroel Orthodox congregation in Berlin, led by Rabbis Azriel Hildesheimer and Ezra Munk.

As mentioned earlier, the initially small Orthodox minority in Munich appointed Heinrich Ehrentreu its spiritual leader in 1885. He never received the title of Rabbi, since legally only one rabbi was permitted there. In 1892, a synagogue, formally a "house of prayer," was built using private member funds. The legal form of this minority was first a "recognized," and later a "registered" association (Ohel Jakob). Josef Perles (1835–1894), the main congregation rabbi at the time and a noted scholar, was a prominent representative of Liberal Judaism. He immediately supported his congregation's executive committee in opposing the Orthodox members' demands for autonomy. In any case, on *Shabbat HaChodesh* 1892, he participated in the consecration of the Orthodox synagogue, when he lit the *Ner Tamid* (eternal light). After the death of Rabbi Perles, Dr. Cossmann Werner of Danzig was appointed the new rabbi of Munich. A brilliant pulpit orator with personal conservative Orthodox leanings, he often argued with the heads of the congregation, who were extreme Neologs. The following is a somewhat humorous dispute. On the eve of the last (eighth) day of Passover, people wanted to say the evening prayer early, before sundown, as on Friday evenings in the main synagogue, and at the same time count the Omer, which is not allowed until after sundown. Rabbi Werner rejected this of course as contrary to the Torah commandments and absurd. At the meeting of the congregation administration to resolve the dispute, my father represented the Orthodox minority on the administration board and was also head of the ritual commission. In view of the irreconcilable differences, he evidently proposed, in all seriousness, to announce the seventh, that is, the previous day of the Omer. The congregation leaders initially agreed, but then realized that they had been made fools of.

Ohel Jakob was initially led by the numismatist Dr. Eugen Merzbacher, a quiet, withdrawn man, respected by all, and definitely averse to conflicts. His reputation was based partly on his father Abraham Merzbacher, an Orthodox scholar in Munich, who had collected a comprehensive and valuable Hebraica and Judaica library. Ohel Jakob was allowed to use the library. After Eugen Merzbacher's death, Dr. Aron Freimann acquired it for the Frankfurt State Library. Professor Gottfried Merzbacher, the famed alpinist and Caucasus and Karakoram researcher, belonged to the same family. Eugen Merzbacher died in 1903, and Sigmund Fraenkel was elected to succeed him as president of Ohel Jakob. This was living evidence of the

[34]For his relationship to the Fraenkel family, see the Fraenkel family tree.**

community's appreciation for Sigmund Fraenkel, given the Feuchtwangers' position as the numerically strongest and financially most prominent family in the synagogue.

Soon after becoming the president of Ohel Jakob, Sigmund Fraenkel began a campaign to support legal safeguards for the Orthodox minority in the Jewish communities in Munich and Nuremberg. His efforts were directed both within the Jewish community as well as outside it, at the Bavarian state assembly (*Landtag*), and at both the Bavarian press and the Frankfurt daily *Frankfurter Zeitung*. For a full decade, he dedicated most of his energy to this campaign, while bearing the brunt of personal attacks and grievances from both the left and right wings. In trying to understand the weight he gave to this task, it should be kept in mind that until then, the Orthodox services in Munich and Nuremberg had no legal basis. They were expressly characterized by the main congregation administrations as "temporarily tolerated." Their spiritual leaders, Dr. Heinrich Ehrentreu in Munich and Dr. Arnold Klein in Nuremberg, were not allowed to use the title of rabbi, nor exercise the corresponding function, give sermons or eulogies, or perform marriages. In addition, the Orthodox also had to pay the full "church taxes" to the main congregation to support the organ, mixed choir, etc., while their own services had to be financed privately.

The following quotation, evidently written by a Christian reporter, from a newspaper article in Munich's *Allgemeine Zeitung*, on September 21, 1907, describes how the situation was viewed at the time: "It might... seem regrettable from an ethical and religious standpoint that it is the state that holds and exercises the ultimate oversight of the internal affairs of the religious communities. State intervention to protect freedom of conscience and prevent oppression of minorities in a religious community would, in any case, be a far better solution than the present situation, in which legal religious bodies, that is, the administrations of the religious congregations, presume to dictate over the conscience of congregation members by recognizing as legal only those religious institutions that correspond to the majority decision."

The struggle for freedom of conscience had not only a practical, but also a fundamental goal. It would either be analogous to Prussian legislation, permitting two religious congregations, Neolog and Orthodox, in the same location and providing the congregations proportional representation and a just tax system not based on arbitrary assessment; or Bavarian Jewry would follow the model of the French law of December 9, 1905, which especially applied to the Catholic Church, but also to Protestants and Jews, implementing separation of church and state.

On this fundamental question, Sigmund Fraenkel fought on three different fronts, two of which were very plausible, but the third presenting a tragicomic puzzle. Naturally, the Neolog majority in the unified congregations of Munich and Nuremberg did not want to give up any of their privileges or jeopardize their autocracy, even through modern voting rights or a system based on state taxes.

Also, Bavaria's legislative (*Landtag*) and executive branches of government were natural opponents of reform. On this issue, Sigmund Fraenkel, although politically among the left-democratic "liberals," also had close interdenominational connections. These included not only the prominent Catholics, such as the *Landtag* president Dr. Georg von Orterer and Georg Freiherr von Hertling, professor of scholastic philosophy at the University of Munich, who was appointed Bavarian minister-president in 1912 and later became German chancellor, but also Protestant lay leaders, such as the bank director Wilhelm Freiherr von Pechmann. On the one hand, the Catholic Center party, which was all-powerful in the Bavarian state parliament from 1903 to 1918, detested the idea of separating church and state and regarded the notion of allowing two Jewish religious communities in one place as dangerous, due to its possible consequences for the Christian denominations. On the other hand, Catholic Church leaders sympathized completely with the struggle against the rape of the Orthodox minorities.

The third front strangely enough consisted of the rabbis in the small Orthodox rural Bavarian communities, such as Ansbach, Regensburg, Aschaffenburg, and Schwabach, among others. They were led by the extremely Orthodox rabbi Dr. Pinchas Kohn of Ansbach, who later ran the World Organization Agudath Israel (Agudah) from Vienna, which even surpassed the Agudah organization of the *yishuv* in its hostility to Zionism. In his position as Executive Director of the World Agudah (1930–1931), Kohn caused quite a sensation when he officially approved of Lord Passfield's White Paper restricting Jewish immigration to Palestine/Land of Israel, and rebuked the *yishuv*'s Agudath Israel for opposing it. On account of this White Paper, Chaim Weizmann announced his resignation as president of the World Zionist Organization and the Jewish Agency, at the Zionist Congress in 1931.

The arguments of this group of rabbis (of small rural Bavarian congregations) against Sigmund Fraenkel's struggle were partly opportunistic. They were used to scare both the small rural congregations with the specter of their splintering, and the taxpayers in the large congregations with having to make an increased sacrifice to the rural congregations. However, the existing legal situation was morally intolerable precisely from the rabbinical perspective. The Christian official responsible for the Jewish religious communities within the Bavarian Ministry for Cultural Affairs was the final authority on matters of Jewish law (halacha), for example whether the *almemar* (bimah) had to be at the center of the synagogue, or a mourner was allowed to recite the Kaddish without putting on tefillin (phylacteries). I remember when this official once called my father on the telephone to ask what the differences were between Orthodox, observant, and traditional Judaism. My father replied sarcastically: "We Bavarian Jews are not sensitive enough to distinguish such nuances. For this kind of answer, you have to ask the same question in Frankfurt am Main."

In reality, people basically feared the views of the Orthodox rabbis and the minority of Orthodox laypeople under their influence. The petition to the Bavarian Ministry for Cultural Affairs, on August 29, 1907, by the Ohel Jakob leadership in Munich, signed by Sigmund Fraenkel and Sigmund Feuchtwanger, and by Adass Yisroel in Nuremberg, signed by Don Sichel and Max Meyer, stated: "In fact, in Prussia it has been shown that only very few congregations have taken advantage of the authority granted by the Secession Law of 1876 to secede from the unified congregation. Instead, the abolition of compulsory congregation membership made the majority feel compelled to show more consideration for the minority. Thus, as a consequence of the Secession Law, the unity of most synagogue congregations was strengthened rather than weakened. The undersigned also think the same about developments in Bavaria. . . ."

That position later provoked extreme attacks from the rabbis of the Bavarian unified congregations as well as the representatives of the Frankfurt religious community (the Secession Community). The monthly *Jüdische Monatshefte* (לעמו דורש טוב), published by Rabbi Dr. Pinchas Kohn in Ansbach, gave the latter a voice. Bavarian Jews claimed that since the Orthodox and the Neologs no longer shared a common faith, it was a sin against the spirit of Judaism to demand from the legislature the option to secede. Members in fact felt taken care of in a unified congregation as long as the needs of the Orthodox were satisfactorily taken into account, as in the main Frankfurt congregation. This led to several paradoxes: in February 1914, the District Rabbi of Aschaffenburg praised the Prussian secession law as the highest achievement of Orthodoxy in a speech in Frankfurt and, barely a month later, bitterly opposed the prospective Bavarian secession law at the general meeting of the Association for the Interests of Observant Judaism in Bavaria.

In the course of his related essays and speeches from 1907 to 1914, Sigmund Fraenkel grappled with this antagonism, most decisively in his 40-page monograph *Austrittsspiel* (The Secession Game) of 1914. A stylistic masterpiece in the form of an open letter to attorney Dr. Isaac Breuer of Frankfurt, it dealt with objective differences, not personal ones. Only a few years later, Sigmund Fraenkel and Isaac Breuer became good friends.

As early as 1907, my father's struggle already had initial unexpected, yet far-reaching, success. The administration of the Munich Jewish community made an official agreement with Ohel Jakob, according to which Ohel Jakob's services and rites were granted legal entitlement, and the Jewish community would contribute an admittedly minimal 2 percent of the religious taxes received to their budget. Also, the rabbi transferred his right to oversee the services to the Orthodox "rabbi assessor," who was also permitted, under certain conditions, to conduct funerals, but not weddings! Somewhat less extensive liberties were attained in Nuremberg. The achievement of legal entitlement was by no means a theoretical quirk. In September 1911, after a women's choir was introduced in the synagogue in Bamberg, the Orthodox minority "illegally" held their own, separate service for the High Holidays. The Neolog rabbi of the Bamberg congregation, in his function as a teacher of religion, punished a secondary school student with a two-hour detention because he had attended the separate services on the Day of Atonement with his father.

Sigmund Fraenkel's large-scale treatise on "The Future Structure of Jewish religious Taxes in Bavaria" is further evidence of his legislative activities for the Bavarian religious communities. This 43-page document was published in 1914 in the *Blätter für Administrative Praxis* (Papers for Administrative Practice) in Munich. His 1911 article, "Die Einführung der Verhältniswahl" (Introduction of the Proportional Vote), referred to all major Jewish communities in Germany. This and other essays also serve to confirm the legal knowledge he had acquired on his own initiative. He suggested that, like the Catholics and Protestants, the Jewish community should be levied taxes in the form of a supplement to the state tax, on a percentage basis. Taxes would no longer be based on assessment by certain members of the congregation administration, who definitely did not have access to reliable data for several thousand taxpayers. This demand, which now seems a matter of course, was ahead of its time. This is all the more apparent when one realizes that, for example, long after the German Revolution, in the 1920s, tax assessment in the Jewish communities of Kurhesse was still carried out by a few community leaders who were not even elected.

At that time, proportional representation seemed a new idea. Sigmund Fraenkel's struggle to implement it in the large religious communities in Germany and to dismantle the existing restrictions in active voting rights with respect to age and citizenship (limited to German citizens) brought great antagonism his way. The urgency of the issue can be illustrated by the example of the largest Jewish community in Germany, the "main congregation" in Berlin, versus the relatively small Adass Yisroel Orthodox secession congregation. In November 1910, roughly 9400 voters participated in the elections, of whom around 5600 voted for the Liberal candidates and about 3800 for the Orthodox ones. Due to the majority rule valid there, the Liberals received all 13 seats, whereas proportional representation would have meant eight seats for the Liberals and five for the Orthodox. In Munich, Frankfurt am Main, Posen, and other congregations with lower absolute numbers, the percentages were similar.

In his 1911 article about proportional representation, Sigmund Fraenkel already opposed the attempt by the Liberals to discredit the Orthodox in the press. They went so far as to place advertisements in the regular dailies, stating that some of the Orthodox candidates were Zionists. "We want to remain German Jews, not 'national Jews,'" they claimed. In 1911, Fraenkel wrote: "Anyone with even the faintest notion of the most elementary principles of a liberal worldview would openly support the conviction that, despite its utopias, the Zionist movement can also claim a seat and voice in the administrative body of our congregation, relative to its followers' numbers. Thus, it must be said clearly and in no uncertain terms, whoever denies this option is on a par with the anti-Semitic campaign against hiring Jews in state and local civil service, and against the election of Jews to honorary municipal and state offices, thereby violating a fundamental principle of a liberal worldview by supporting a law of exception."

Whereas, in 1911, the "Zionist question" was mainly part of Sigmund Fraenkel's polemics against the Liberals, a decade later, given the changed political circumstances in the German republic, his stance opposed the majority of Bavarian Orthodoxy. In January 1921, in the Munich weekly *Das Jüdische Echo* (The Jewish Echo), he published an article on "why Orthodox and Zionists should be allies in the upcoming Jewish Community elections." By then, after ongoing struggles since 1907, the three main constitutional goals, secession option, proportional representation, and central taxation authority, had been achieved. The upcoming elections were supposed to invigorate the Association of Bavarian Jewish Communities and, at the same time, provide the structure for reestablishing congregation life, as was initiated by Alfred Neumeyer, chief state judge of the Higher State Court and later highly deserving president of the Munich congregation. It had become clear that the old antagonisms relating to coerced affiliation with a congregation, which arose because of the Neolog postulates for synagogue services, such as an organ, a mixed choir (including Christian women), and a three-year Torah cycle, were obsolete. Problems of a general nature, pertaining to more than just the synagogue itself, had to be tackled. Most important were those issues involving foreigners, the exclusion from passive electoral rights until then for Jews not naturalized in Germany, and the establishment of a Jewish elementary school. Since the Orthodox and Zionist stances on these points concurred, joining together for the elections made common sense. This did in fact occur.

In the 1914 elections to the Knessia Gedolah (Great Congress) of the recently founded Agudath Israel, the overwhelming majority of German Orthodox Jews refused to list Orthodox Jews with Zionist leanings as candidates. These elections were postponed when war broke out. Similarly, their coalition for the 1921 election mobilized the Orthodox anti-Zionists. Sigmund Fraenkel defended himself within his close circle by explaining it as a tactical, and not an ideological, alliance. By contrast to the first years of the century, significant circles among the German Orthodoxy tended at the time—unhindered, and even strengthened, by the war and revolution—to make more radical and fanatical decisions.

December 26, 1907, was an important date for reconciliation and bridging differences. Dr. Salomon Breuer, rabbi of the Jewish Community (Israelite Religious Society, IRG) in Frankfurt am Main, who was the son-in-law of Samson Raphael Hirsch, the first rabbi in that secession community, led the Free Association for the Interests of Orthodox Judaism in Germany (*Freie Vereinigung für die Interessen des Orthodoxen Judentums*). He took an extremely Orthodox course, which remained limited to a narrow circle, largely in Frankfurt and the environs. On that winter day in 1907, at a conference in Berlin, representatives of congregational Orthodoxy[35] actively joined the Free Association for the first time. They included leading economists, such as Louis Feist of Frankfurt, Benjamin Hirsch of Halberstadt, and Sigmund Fraenkel. They considered the rabbis of Frankfurt, Marcus Horovitz (until 1910), Nehemia Anton Nobel (until 1922), and others, including Selig Auerbach (who died in 1901), their spiritual leaders. At the Berlin conference, Rabbi Breuer indicated, with respect to the newly won comrades, "Even though on one point, regarding the secession issue, their accord is of a tactical, political nature, on which they did not previously agree with the leaders of the Free Association, their entire thinking, feeling, and desires, in fact, agree with those of its early founders." However, as early as 1912, when Rabbi Dr. Pinchas Kohn of Ansbach and Dr. Isaac Breuer of Frankfurt founded the monthly publication *Jüdische*

[35]"Congregational Orthodoxy" refers to those willing to remain in the unified congregation with the Liberals, as long as they had their own synagogue, in contrast to "Secession Orthodoxy." The secession congregation led the "Free Association."**

Monatshefte, the extreme, polemic tendencies of the Frankfurt-led Orthodoxy were renewed and intensified. They began with harsh attacks against Jacob Rosenheim, lay leader of the German Agudists, who advocated secession, declaring it a question of "denomination." They claimed that the Bavarian Orthodox had no right to revise the law, as long as Orthodox and Neolog Judaism were not declared "separate denominations."

Also, after the abolition of forced membership in the community in 1918, Sigmund Fraenkel fought against secession from the main Jewish community. In cooperation with the head of the congregation Dr. Neumeyer, he consolidated all the Jews in Bavaria in the Association of Jewish Communities (Verband Israelitischer Gemeinden). On his deathbed in September 1925, he summoned the leaders of the Munich Jewish community, Dr. Neumeyer and Dr. Elias Straus, and struggled to utter the following words: "The Munich Jewish community should remain a unified congregation."[36]

Despite the alliance with the Zionists for the election, Sigmund Fraenkel was never a Zionist and never managed to fulfill his wish to visit the Land of Israel. Nevertheless, his views on Zionism differed from those of the official Orthodoxy. When his sons were close to their university qualification examinations (*Abitur*) and starting their studies, between 1905 and 1909, he would tell the northern German students invited to spend Friday evenings with us, much to their consternation: "I have nothing against my sons joining a Zionist student organization, in their early semesters." In response to the occasional question about why in their early semesters only, he would reply, "Meanwhile, they will recognize how impractical Zionism is." In fact, my mother raised objections when I joined the Zionist Organization in 1915, as a soldier at war. But neither my father nor my teacher, Rabbi Ehrentreu, opposed my decision.

Regarding Sigmund Fraenkel's contribution to German-Jewish "foreign policy," before the First World War there is nothing particular to report, since at least in Bavaria one did not experience much aggressive anti-Semitism. Also, my father was not all that interested in combating secondary affronts, such as the virtual exclusion of Jews from the Prussian officers' corps and from full university professorships. Such issues were taken on by the Central Association of German Citizens of Jewish Faith and the Society for Combating Anti-Semitism. The first such event of any significance, relating to Jews from Eastern Europe, and not from Germany, emerged during the war.

Already in late 1914, Sigmund Fraenkel, who was in his mid-50s, was sent by the German government into the war zone of the German-occupied French textile centers of Lille, Roubaix, and other cities in northern France. As an expert in raw wool, he was asked to assess the large stocks there and provide suggestions on how to utilize them.

[36]See *Bayrische Israelitische Gemeindezeitung* (1925), no. 10; and *Jüdisches Wochenblatt* (Frankfurt am Main, 1926), no. 38.

The second time he was in a war zone, it was on the eastern front, and not for economic, but for Jewish interests. By conquering large parts of Congress Poland, Lithuania, and adjacent parts of Russia (almost all of which were parts of the Russian Pale of Settlement), the Germans were brought into contact with hundreds of thousands of Jews. This posed a great problem for the military authorities, since they did not usually display much understanding or empathy. Just as different from the Western, assimilated German Jews as from Slavic people, and ostensibly "German-speaking" (as native Yiddish-speaking Jews believed they were), most of them were strictly observant, and some were left-wing socialists. To the German officers, neither harshness nor understanding seemed to be very effective with such people. The satiric books of Sammy Gronemann[37] portray a fitting picture of the contrasts and misunderstandings at that time.

In late August 1914, the German Committee for the Liberation of Russian Jews was formed in Breslau; in early 1915 it was expanded as the Committee for the East. Information on the committee's organization and activities are given in S. Adler-Rudel's essay, "East-European Jewish Workers in Germany."[38] However, rather than deal with this general committee, consisting of various anti-Zionist and Zionist circles, the German government preferred to work with the Free Association for the Interests of Orthodox Judaism in Germany. In February 1916, the association sent Rabbi Dr. Pinchas Kohn of Ansbach and Rabbi Dr. Emanuel Carlebach of Cologne (older brother of Dr. Joseph Carlebach, who later became the outstanding Chief Rabbi of Altona and Hamburg) to serve as representatives in Warsaw, where they remained until the end of the war. Despite their close connections to the influential "Gerer Rebbe" (A.M. Alter), the two rabbis encountered little affection or trust from the Polish Jews. This was due not only to the rabbis' Agudistic and anti-Zionist views, which were criticized even by the German Agudah, but also because of their over-identification with the German authorities and their worldview.[39]

However, the German occupation was not limited to Congress Poland, the part of Poland ceded to Russia at the Congress of Vienna in 1815. It also included additional, mostly northern and northeastern areas of Russia such as Grodno, Kovno, Vilna, and other Lithuanian cities, where hundreds of thousands of Jews lived. Since there was a German civil administration within the Polish General

---

[37] *Towuwabohu* (Berlin: Welt-Verlag, 1920); *Hawdoloh und Zapfenstreich* (Berlin: Jüdischer Verlag, 1924).

[38] Appeared in the *Leo Baeck Institute [of Jews from Germany] Yearbook* 2(1): 136–165 (London, 1957). It includes an interesting portrayal of the further developments, in Poland. See also S. Adler-Rudel, *Ostjuden in Deutschland 1880–1940.* Schriftenreihe wissenschaftlicher Abhandlungen des Leo Baeck Instituts (Tübingen, 1959).

[39] See the essay by Alexander Carlebach, Emanuel Carlebach's son: "A German Rabbi Goes East," in the *Leo Baeck Institute [of Jews from Germany] Yearbook* 6(1): 60–121 (London, 1961). The article primarily contains a series of letters written by Rabbi Emanuel Carlebach in Warsaw to his family in Cologne, from January 18, 1916, to November 20, 1918. Suffice it to cite from the letter of October 28, 1918: "As you can believe, after all my conservative leanings, I do not feel undivided joy regarding the democratization of the Reich." The second letter in the series is dated: "Warsaw, Emperor's birthday, 1916"!

Government, but a military one outside it, known as Oberkommando Ost (Supreme Commander East), or "Ober Ost," it was incomparably more difficult to influence the lot of those Jews. It is not surprising that the Committee for the East, which even in Poland ranked behind the Free Association, initially failed to gain access to the military administration.

This is where my father stepped in. Coming from Munich, he took advantage of the fact that Prince Leopold of Bavaria was the Commander-in-Chief in the East. On September 18, 1916, he contacted Bavarian Foreign Minister Georg Count von Hertling,[40] whom he knew, asking him if the Commander-in-Chief in the East would receive a delegation of four (specifically named) leading Orthodox German Jews, to review their proposals to appease and inform Lithuanian Jews and work out a suitable relief operation by German Jews to benefit their Lithuanian brothers. The army offices struck out two of the suggested names, the head of the Free Association and the rabbi of the Secession Congregation in Berlin. The other two, Sigmund Fraenkel and his cousin Dr. Abraham Hirsch of Halberstadt, who was Benjamin Hirsch's son, were received by General Field Marshal Prince Leopold in the headquarters in Brest-Litovsk and by the "Senior Quartermaster" in Bialystok in April 1917.

The success of this trip was immediately apparent. On the recommendation of Sigmund Fraenkel, Hermann Struck, famous artist and religiously observant Zionist, was made director of a new department for Jewish Affairs established under the Commander-in-Chief. Struck already had connections with the Committee for the East. The books by Sammy Gronemann, mentioned previously, describe the department's activities and the interesting staff made up of Jewish soldiers. Although a polemic between Orthodox, Zionists, and Liberals of course unfolded in the Jewish press in Germany regarding the department's activities and aims, considerable sums were collected for the Lithuanian Jews, and tensions between them and the military authorities markedly declined.

In 1919, long before it gained a stronghold in northern and western Germany, aggressive anti-Semitism began in Bavaria, especially in Munich, the Nazi home turf. To understand the developments, it should be noted that after the German defeat, and the German Revolution in November 1918, left-wing radicals, including some Jews, initially came to power in Munich. In March and April 1919, as these groups became increasingly bolshevized, street fighting began. To give them something in hand against the advancing government troops from northern Germany, some respected citizens were taken hostage.

[40]In September 1915, the same Baron (later Count) von Hertling had already given the Free Association delegation a written introduction to the Governor General of Poland, General Hans Hartwig von Beseler. Could the shared interests of the Catholics (with Hertling as the leader of the Catholic Center Party) and Jewish Orthodoxy possibly explain the success achieved by the Free Association with the Committee for the East? The relevant original letters, including one dated March 19, 1917, from the Commander-in-Chief in the East to Minister von Hertling, are reprinted in *Sigmund Fraenkel* (see note 20), pp. 209–215.

Sigmund Fraenkel anticipated that Jewish participation in the radical leadership would have a disastrous impact on Bavarian Jewry. With agreement of the editorial offices of the leading Munich daily, *Münchener Neueste Nachrichten* (*MNN*), he wrote what was, in retrospect, a somewhat prophetic three-page "Open Letter to the Gentlemen Erich Mühsam, Dr. Arnold Wadler, Dr. Otto Neurath, Ernst Toller, and Gustav Landauer," on April 6, 1919. On the morning of April 7, when Sigmund Fraenkel attempted to deliver the open letter to the *MNN* editorial offices, they were already occupied by the Red Guards. Fortunately for him, it was not published at that time, since he might have otherwise shared the same fate as the hostages, who were murdered in the Luitpold Gymnasium (academic secondary school) in Munich in late April, after several Jewish leaders, particularly Landauer, resigned from the leadership.

In the following years, despite his awareness of the danger, Sigmund Fraenkel continued to write articles against the anti-Semitic attacks by the Nazi movement, which increased rapidly in Munich. In particular, he objected to demands to prohibit kosher slaughter (*shechita*) in Bavaria, which were based on blatant lies. An article he wrote in June 1921 in a nonlocal paper, the *Berliner Tageblatt*, exposing the methods of Munich police chief Ernst Pöhner, was especially held against him by the Bavarian press. Pöhner (who in November 1923 was a coconspirator of Hitler in the Beer Hall Putsch) defended the riots of the nationalist German-*Völkisch* movement and the hate propaganda in the Nazi paper, *Der Völkische Beobachter*. In that article, my father warned the Bavarian state government of the consequences of its pussyfooting. Among other things, he reported on the Catholic cleric in Munich's Holy Ghost Church who was interrupted while speaking about Jesus, with the jeering remark, "How can Jesus help us if he himself was a Jew!"

Thus, it did not come as a surprise when on June 22, 1923, there was an assassination attempt on Sigmund Fraenkel. As reported in the far-right newspaper *Münchener Neuesten Nachrichten*: "The vice president of the Munich Chamber of Commerce, Councilor of Commerce Sigmund Fraenkel, highly esteemed among broad circles in Munich and Bavaria for his economic and charitable activities, was brutally attacked in the streetcar by some rowdies on Thursday evening while traveling home from a trip ... and beaten bloody as he changed trams. With the words 'You Eastern Jew, you Jew swine, get out with your stolen luggage,' he was punched in the face and then hit over the head with a hard object. In front of the building where Mr. Fraenkel lives, another three people who had been waiting for him raced out, yelling that, were he not such an old man, they would have beaten him to death. When reprimanded by the policeman accompanying Mr. Fraenkel, the rowdies responded, 'What is the big deal, one Jew less?'"

The press outside Bavaria and Germany called a spade a spade, reporting that Nazis wearing swastikas had beaten Sigmund Fraenkel with rubber clubs. The perpetrators got off with light penalties. Citing another article, "The Jew Fraenkel and Rabble-Rousing against Hitler in Bavaria,"[41] in the *Völkische Kurier*, on March

[41]Original German title: "Der Jude Fraenkel und die Hitlerhetze in Bayern."

14, 1925, which temporarily replaced the banned *Völkischer Beobachter*: "A few weeks ago, Councilor of Commerce Sigmund Fraenkel, an only too well-known Hebrew and 'fellow citizen' of our native city, tried to incite the tourism association with the same arguments against Hitler. . . . This industrious businessman. . . is the soul behind all the rabble-rousing against Hitler. . . . The Jews of the Old and the New Worlds had a rendezvous in Upper Bavaria. Jewish scheming has spread between Munich and Garmisch, so at times you might think that our beautiful homeland has already been completely bartered away to the district offices of Jerusalem. And all this at a time when Hitler was at the height of his power!"[42]

Sigmund Fraenkel was a healthy, strong, and unshakably optimistic man. However, during the First World War, which from early 1917 on he viewed as a lost cause for Germany, he became a different person. His concern for me also contributed to this change, since from September 1914 until the end of the war, I was a soldier in the field, and, in the summer of 1917, I hovered between life and death for several weeks. However, his virtually prophetic assessment of the future economic and domestic policy in Germany weighed even more heavily. Like almost all German Jews of his time, he viewed Germany as a rising world power. He especially feared the backlash of such developments on the fate of German Jewry. A Cassandra prophecy, it was fated to be correct yet not heeded, since, until the summer of 1918, the overwhelming majority of Germans, including German Jews, fully expected a German victory and did not even consider the consequences for Europe of German domination.

In contrast, the decline in my father's personal prosperity barely seemed to disturb him. A wealthy man before the war, he lost much of his fortune in the galloping inflation from 1918 to 1923. In addition, he was absolutely meticulous in paying asset-related and other taxes after the war, although the rest of the business world proved to be rather lax in that regard. It was not merely a moral obligation for him, but arose from a desire to maintain unassailable integrity in his dual prominent positions as a government advisor in economic affairs and leader of the Bavarian Jewish community.

Certainly, cheerful events after the war lightened his mood. Once again his old self, he wrote and edited high-spirited and optimistic marriage poems looking forward to a better future, in particular, to the engagements and marriages of his three older children. He was already seriously ill at the wedding of his second daughter in 1924. His youngest daughter did not get married until 1934, in Jerusalem. The modest villa in Bad Reichenhall, which he bought in 1917, was also a source of joy to him. Every summer, my father and mother hosted their children and grandchildren there. He worked there during the vacations, attended the concerts of the spa orchestra as a local, and entertained many friends.

[42]In late 1924, after serving a short sentence following the Beer Hall Putsch in November 1923, Hitler (who lived very near to us, on Thierschstrasse no. 41, by the way) was again favorably received by Bavarian Minister-President Heinrich Held, leader of the Catholic Bavarian People's Party.

However, the political events of 1921–1923, especially the assault in 1923, shattered his health, although this did not become clear until March 1924. On a business trip to London, my father suffered severe angina pectoris. Despite the care of eminent physicians, his heart, which had been weakened over the previous few years, never regained its resilience. In the summer of 1924 he recovered again temporarily, but had to increasingly limit his activities. From the summer of 1925 on, he was bed-ridden and his strength continually declined, which must have been particularly painful for such an active man. On September 19, 1925, around *Rosh Hashanah*, his heart began to fail again. He summoned the leader of the Munich Jewish community to his bedside to express his final words and wishes, and requested forgiveness from groups and individuals for any personal hurt inflicted in the heat of the moment. He strongly encouraged his children to adhere to the Jewish tradition and to study Torah. While his body declined, his mental acuity remained sharp. He said to me: "I do not complain about dying relatively early; the shortness of my life is balanced by the fact that it was full and successful, and I experienced much happiness." The close of the Day of Atonement, on Monday, September 28, he remarked: "You will have a short shivah." He assumed his funeral would take place before the *Sukkot* festival that started on Friday evening. On Wednesday afternoon, September 30, his death throes ended. On Thursday afternoon, he was buried in the old (Thalkirchener) Jewish cemetery. Among the mourners at the cemetery were people from his business circles, including the Bavarian Minister of Trade and Industry Wilhelm von Meinel and the vice president of the Munich Chamber of Commerce, Privy Councilor Gautsch.

Because obituaries are forbidden by the Jewish religion between *Yom Kippur* and *Sukkot*, there were only short eulogies at the cemetery. After the holidays, a memorial gathering took place in the Ohel Jakob synagogue. Among the eulogies, the speech given by Dr. Elias Straus[43] at the Jewish community's memorial service, which took place on October 27, 1925, stands out. In accordance with his express provisions and those of his grandfather B. H. Auerbach, in addition to his name and the dates of birth and death, his gravestone only has the words: מגזע גדולי ישראל.[44]

I reported on my father's life and works in such detail since it represents an important segment in the history of Bavarian Jewry spanning half a century. I can be somewhat briefer regarding my mother Charlotte (Chaya Sara), daughter of Isidor Neuburger and Rosa Neuburger, née Fraenkel, born in Munich on the 4th of Tammuz (June 24) 1868. This reserve is not only due to the fact that, at the time of writing, my mother is still alive and has just celebrated her 97th birthday.[45] Rather, the main reason lies in my mother's nature and lifestyle, which can best be characterized by the much-quoted verse from Psalm 45:13: "The king's daughter is all glorious within." The intimate atmosphere of the home was her realm, and,

[43]Printed in the *Bayrischen Israelitischen Gemeindezeitung* (1925), no. 10, and in the commemorative book *Sigmund Fraenkel*, pp. 329–341 (see note 20).

[44]"Of prominent Jewish ancestry".

[45]Charlotte (Chaya Sara) Fraenkel died on October 24, 1965.

first and foremost, her concerns were for her spouse, parents-in-law (whom she knew well from an early age since they were close relatives), and her own children. Typical statements and personal reflections from her life cannot be reproduced because they are too intimate. These include her farewell words when I went to fight in the war, in 1914; and her heroic return to a demanding daily routine after her oldest daughter, mother of four young children, was torn from her by a severe illness when she was only 30 years old, in 1927.

When I think back on my childhood, my mind is of course filled with countless memories of my mother. I shall mention just a few very characteristic traits. She was profoundly religious throughout her life. Her Jewish education, which was relatively extensive for Munich at that time, especially in the *Siddur* and *Chumash* [Pentateuch], enabled her, after emigrating to Palestine/Land of Israel in 1934, to quickly gain sufficient proficiency in modern Hebrew to converse with her grandchildren and great-grandchildren. She was competent in English and French, even at an advanced age. She also had proverbial modesty and an undemanding nature, making do with the most simple residence,[46] clothing, and food, while passing on the "good things" to her husband and children. She showed absolute subordination to her in-laws, in particular to her rather difficult mother-in-law and to her husband. She disliked social obligations and wished to remain in the background. In childrearing, she artfully demanded obedience while bestowing loving care.

Her physical strength and mobility gradually waned after her 90th birthday, but she remained mentally alert; and a regular and elegant letter-writer, just like her mother, who had been a true artist in that regard. Not only her children, children-in-law, and grandchildren, but also nephews and nieces and distant relatives, share their experiences and problems with her. As the last survivor of her generation, she continues to represent a focal point for the younger family members.

This chapter is about the decades during which Munich grew from being a provincial regional capital to a center of German intellectual life, especially in the areas of art and literature. Readers might ask: Why is there no mention at all of this in the first section and only very little in the second? The answer is because these are the memoirs of an Orthodox Jewish family. Whereas some families from Liberal Jewish circles were active in Munich's intellectual life, in my parents' generation, my uncle Dr. August Feuchtwanger was perhaps the only religiously observant Jew who internalized the Munich atmosphere, especially in the musical sphere. In my own generation, many Jews from Ohel Jakob did so, such as the children of Sigmund Feuchtwanger, especially Lion and Ludwig (known as Ludschi). Of course, these men and women, some of whom were highly talented, soon distanced themselves from the Jewish tradition.

---

[46]Not until 1915 did my parents move out of the simple apartment on Klenzestrasse into a beautiful residence in a nicer location at Thierschstrasse no. 25. A built-in sukkah was erected there.

Chapter 2
Childhood and Adolescence in Munich (1891–1910)

It is as if my life unfolded in 19-year cycles. After the first such period as a child and teenager in my parents' home, I spent the next 19 years (1910–1929) as a student, and then as a professor at German universities, admittedly also including four years as a soldier in the First World War. During the next 19 years (1929–1948), I was a professor at the Hebrew University in Jerusalem, in the Land of Israel/Palestine under the British Mandate. Although I continued as professor after 1948, and as emeritus professor from 1959, the fourth 19-year period focused more on activities for the benefit of the State of Israel rather than strictly on my academic career.

In any case, because of my fascination with the Hebrew calendar, the number 19 is special for me.[1] I was born[2] on 9th of Adar Rishon (the First Adar) 5651, which was February 17, 1891. It was a leap month of a leap year in the Jewish calendar, in the eighth year of the 19-year cycle, in which the Jewish holidays fall on the latest Gregorian dates. In 1948, three 19-year cycles after my birth, the "lateness" of the Jewish holidays played a significant role in the Israeli War of Independence.

I can place my earliest conscious impressions rather precisely, namely, to when I was three years old. Shortly before my fifth birthday, I began reading Hebrew and soon learned to translate the Pentateuch. I did the translations rather mechanically,

[1]The normal Jewish calendar year consisting of 12 lunar months is roughly 11 days shorter than the solar year of 365 days (except in a solar leap year, which has an extra day). The leap month Adar Rishon (The First Adar) is added in seven years (the 3rd, 6th, 8th, 11th, 14th, 17th, and 19th) of each 19-year cycle of Jewish years so that the seasons in the lunar year keep pace with the solar ones. In leap years, the Jewish holidays occurring after the leap month fall relatively late in the Gregorian solar calendar. In the eighth year of each 19-year cycle, they fall on the latest Gregorian date, since the largest number (5) of leap months have been added in the preceding 11 years.**

[2]My "official" name was Adolf and my "Jewish" name, Abraham Halevi. In the 1930s I dropped the "Adolf" for understandable reasons and replaced it with my Jewish name when I became naturalized in Palestine/Land of Israel.

A.A. Fraenkel, *Recollections of a Jewish Mathematician in Germany*,
DOI 10.1007/978-3-319-30847-0_2

relying on Rudolf Fuchs' translation.[3] However, contrary to the widespread view, this did not do me any harm. I learned Hebrew quickly and effortlessly. I came to appreciate it as a literary language, which was not distorted by error [as in spoken language]. In addition, translating difficult or rare Hebrew words into German enabled me to learn their precise meaning earlier than I might otherwise have done (in contrast to children who spoke Hebrew as their mother tongue). Indeed, through the Pentateuch translation I even learned some German words I was not familiar with as a child. I can recall the examples: "*zurechtweisen*" (reprove) (Genesis 21:25) and "*Pfosten*" (doorpost) (Exodus 12:7), not to mention animal and plant names, among other realia, which I only got to know through the Bible.

In my first two years of Hebrew instruction, I had four different teachers. The name of one of them, Benzion, greatly impressed me.[4] He was a knowledgeable Sephardic man from Serbia—Sarajevo, I believe. At that time, there were hardly any Hebrew teachers in Munich, and the few who did teach there moved on after a short time. But from 1897 to 1898 on, my brother Eugen Joël and I were fortunate to have an outstanding long-term teacher, the unforgettable Leopold Rotter. Educated at the Pressburg (present-day Bratislava) Yeshiva, Rotter was hired as a *chazan* (Jewish prayer leader) and teacher by the Ohel Jakob congregation, and as a *shochet* (Jewish ritual slaughterer of cattle and fowl) by the Jewish religious community. After about six years, he was pleased to accept a teaching position at Berlin's Adass Yisroel congregation, so he could stop working as a *shochet*. My brother and I moved heaven and earth to keep him from moving away. Ultimately, an arrangement was found to make it possible for Rotter to earn his livelihood as a teacher and *chazan* only, though it involved great sacrifices by both Ohel Jakob and our family.

My father's interest in our Hebrew-Jewish education was not expressed solely through his concern for our private instruction by Herr Rotter, and, later by Rabbi Ehrentreu. During the winter, every Friday evening after coming back from synagogue, before dinner, he studied Rashi's commentary on the current weekly Torah portion with us. Thus we also became familiar with Aggadic *Midrashim*. In view of my later development, my father's skipping over some Rashi passages, with the comment, "That is just grammatical," was paradoxical. Hardly a decade later, and increasingly throughout my life, I became enormously interested in Hebrew grammar, viewing those Rashi remarks as the pearls of his Pentateuch commentary.

I was born in February and, according to the municipal regulations, I was not permitted to start school when I was only five and a half years old. The school year began in September, and only children who had reached or were close to their sixth birthdays could register for school. My ambitious father circumvented this by arranging private tutoring in arithmetic and German reading and writing, from September 1896, which was some time after I started my Hebrew lessons. As a

[3]Rudolf Fuchs, *Die Thora und die Sprache* (complete text with German translation on corresponding lines) (Vienna: k.-k. Schulbücher Verlag, 1878–1899; early editions of the five books of the Pentateuch).**

[4]Benzion means "son of Zion."**

present for my sixth birthday, I clearly remember my tutor letting me choose my favorite of three subjects for my lesson that day. I loved reading and hated writing, also in Hebrew, but I was too embarrassed to suggest reading, so I chose arithmetic. The tutoring enabled me to enter school directly into the second grade at six and a half years old. In retrospect, my father had shown great foresight, since the First World War started only six months after I completed my doctorate in January 1914.[5]

Primary school was more democratic in Bavaria than in Prussia. In Prussia, parents who wanted their children to get a secondary school education and could afford it would send them for the first three years, from the ages of six to nine, to a preparatory school (*Vorschule*) affiliated with the secondary school. This three-year education was equivalent to four grades in the primary school (*Volksschule*) for "ordinary" children. The children would then attend nine grades in secondary school, which in Bavaria was called middle school (*Mittelschule*). Bavaria did not make any such distinctions. In the first four years of primary school, a cabinet member's son might sit next to a laborer's son. Thus, there were 13 years of schooling until the university entrance qualification examination (*Abitur*), which students normally completed when they were 19.

Since I entered second grade when I was six and a half, I was only at primary school for three years. At that time, Munich's primary schools were denominational, although the municipal administration was largely liberal. The Catholic Center party had the greatest influence in the state parliament and government in Bavaria, so the great majority of over 30 public primary schools was Catholic; only a minority of the schools were Protestant. There were only two so-called "simultaneous" schools, attended by both Catholic and Protestant students. By chance, one of them, the Klenzestrasse School [for boys], was only two minutes from the house where our family lived from 1895 to 1915, so both my brother and I went there. At that time, there were no coeducational primary or secondary schools.

There were 63 students in my class, mostly from the lower social classes, since the school was in a working-class neighborhood. I still clearly remember two of my teachers. The only noteworthy detail was the trend among the largely "liberal" teachers to teach us about drinking beer. At that time, at least in Old Bavaria, starting beer consumption at a tender age was not only considered pleasurable, but virtually a health requirement. Even in Jewish households, it was natural for the father and his sons to down half-liter glasses of beer at lunch and dinner. Beer drinking was a national and almost Catholic-religious pastime. The Bavarian State Parliament, led by its president, would make an annual pilgrimage to the Nockherberg, a terraced beer garden at the Paulaner brewery, to tap the Salvator (a dark Märzen, or March beer) barrel. Drunks were often seen on certain streets in Munich, especially during Fasching (carnival). This explains why Bavarian physi-

[5]Franz Ferdinand was killed on June 29, 1914. Germany declared war on France on August 3, 1914.**

cians were well-acquainted with the "beer-drinker's heart,"[6] which was much enlarged in comparison to a normal heart.

It had been impressed upon the teaching staff of the city of Munich, a majority of whom were politically leftist, that is, liberal and not social democratic, to warn adolescents of the dangers of beer-drinking. I recall my brother politely pointing out at lunch: "Papa, if you keep on drinking so much beer, it'll make you dumb. That's what our teacher told us." My father's reply: "I have been drinking for so long and I'm not dumb," was followed by brother's retort: "Just wait and see. You'll notice it later!" Indeed, neither my brother nor I ever drank beer, not even in beer cellars.

The following anecdote is pertinent here. Until the 1880s, Munich's inferior drinking water was notorious due to the endemic spread of typhoid fever. Once the aqueduct was built, the excellent water from the Mangfall valley was brought to the city. Munich's water is still distinguished among water systems of major European cities, at least up to my last visit to Munich, in 1932. Ten years after "good" water had come to Munich, a Berliner, who was traveling in the fast train from Berlin to Munich, commented to his travel companion from Munich: "I heard that you have had first-rate water for quite a while already," to which the Munich man responded in Bavarian dialect: "Yeah, I heard that too."

During my primary and secondary schools years, I did not really enjoy much camaraderie, and certainly no real friendships. This was my own fault. Until my marriage, I essentially remained a loner and recluse. As much as religion played a role in this, it was not my being Jewish but rather Orthodoxy that hindered my social activity, which was difficult even with my Jewish schoolmates. Many of them were so assimilated that, at their bar mitzvah celebrations, which I was compelled to attend, they would tell anti-Semitic jokes, probably without fully understanding them.

Catholic students sometimes called me or spoke about me as a "Jew pig," which in and of itself was not so bad, but was part of the crude Bavarian parlance, of which even I was not totally free. Such crudeness later sometimes led me to cause offence when I was a student at northern German universities, especially among Jewish fellow students. At the time, true anti-Semitism was uncommon in Catholic Bavaria and Württemberg-Baden, in contrast to Prussia and Austria, where the Jews had a greater presence, both in relative and absolute terms. In states where Catholics were numerically in the minority, they generally avoided expressing animosity toward Jews. Especially from the time of Bismarck's *Kulturkampf* ("culture war") until the end of the German Empire after the First World War, German Catholics were often made to feel their minority position. The situation was completely different in the Federal Republic of Germany after the end of the Second World War, when, after Prussia's defeat, Bavaria and the Rhineland with their Catholic majorities dominated.

[6]Alcoholic cardiomyopathy, a disease associated with alcohol abuse, is characterized by enlargement of the heart and low cardiac output. *Mosby's Medical Dictionary*, 8th ed. (Amsterdam: Elsevier, 2009).**

In September 1900, I started attending the humanities-oriented Luitpold Gymnasium (academic secondary school) on Müllerstrasse in our neighborhood. I spent nine happy years there. Along with the influence of my parental home and my private Jewish studies, those years at secondary school certainly had the greatest influence on my development, even though the main subjects in school were Latin (which I had for nine years) and Greek (six years), whereas mathematics, which would later become my profession, was taught only to a modest extent. The mathematics curriculum, which was very solid, was taught by an experienced, albeit rather untalented, secondary-school teacher (*Studienrat*).

In 1900, our school had a liberal atmosphere, thanks to the principal, Dr. Wolfgang Markhauser, who distinguished himself through his great kindness. In the summer of 1902, he reached retirement age and was replaced by the "uncrowned king of Bavaria," Dr. Georg von Orterer, who was later conferred noble status and the title "His Excellency." Head of the Center Party and longtime president of the Bavarian State Parliament, he was transferred from the world-forsaken town of Eichstätt to the Bavarian capital for political reasons. Like all gymnasium principals, he was a classical philologist. He confirmed his reputation for old-fashioned strictness by giving instructions that even for the Prussia of the time sounded ridiculous, even unbelievable. For example, students had to be home by 7 PM unless they had special permission or were accompanied by their parents. He sometimes sent the school caretaker into the buildings to check whether certain students were there. While Munich was known especially for opera, it also had very good theaters, which we were only allowed to attend with the principal's permission. When I was 17, I requested to go to Lessing's *Nathan the Wise*. The well-meaning von Orterer responded, in the Catholic Templar style, "No, go to an irreproachable play instead!" One 19-year-old student was forbidden to go and see Schiller's *The Robbers*.

The rush to attend the Luitpold Gymnasium declined once Orterer became principal. However, I always got along well with him, especially when he became my Latin teacher in my final year, since the plenary session of the Bavarian State Parliament only convened in alternate years, and it was not in session that year. Von Orterer's courteous behavior to me was certainly due to my affinity for the classical subjects and knowledge of the Bible, but also because he liked all kinds of "orthodoxy." When I asked for permission to attend services on Hoshana Rabbah, the last day of *Sukkot*, which I explained was the "highest Jewish semi-holiday (*Chol HaMo'ed*)," I was immediately permitted to come to class late.

I respected the academic secondary school teachers in Bavaria, especially the homeroom teachers, that is, the main class teachers, who usually taught Latin, Greek, and German. However, German and modern German literature, but not German composition, had only secondary status as compared with classical languages and literatures. The teachers of these subjects all had good university training but were extremely poorly paid, with starting salaries of less than 3000 marks a year. Consequently, they lived frugally, and the ambition of many was to save up enough over two decades to be able to afford a modest Easter or summer vacation in Greece. Some were so devoted to their subjects that their enthusiasm for classical antiquity was contagious to gifted students. Among them, the robust Dr. Adalbert Ipfelkofer had a somewhat military style. I did not like him very much as a person, but, in sixth grade, he knew how to arouse my enthusiasm for Homer. In the higher classes, the sensitive and all too delicate Dr. Eduard Ströbel was the perfect teacher of Plato. Plato and Homer were my favorite reading materials in school; the latter made me greatly appreciate the sound and rhythm of the language. I was also fascinated by the translation of modern German texts into idiomatic Ciceronic Latin. My penchant for Homer and Plato needs no explanation, but I would like to explain why I enjoyed the translations. A mother tongue is best grasped through the medium of a foreign language. We are seldom consciously aware of its concepts and grammatical constructions, and often can only understand and analyze them when they are translated into an idiom that is as distant as possible from the native language. This is a kind of forced defining process. It might explain why academic and regular secondary school students in Israel have unsatisfactory competence in their Hebrew mother tongue. Translating Hebrew texts into idiomatic English or French, and other languages such as Latin and Arabic, has unfortunately been removed from the curriculum for practical reasons. However, such translation would greatly facilitate a good and eloquent Hebrew style. In addition, translations, especially into Latin, which promote analytical thinking and the ability to define things as precisely and succinctly as possible, are also useful for other disciplines, such as mathematics.

Here I have been referring to humanities-oriented academic secondary schools, of which there were five in Munich around the turn of the twentieth century. By contrast, there was only one secondary school (*Realgymnasium*), focusing on mathematics and modern languages, where Greek was not taught. However, modern schools, especially natural science-oriented secondary schools (*Oberrealschulen*), did not become popular in Bavaria until the second decade of the twentieth century. Girls aspiring to a "higher education" attended girls' secondary schools (*höhere Töchterschulen*).

French was the only required modern foreign language. Starting in the sixth grade in the academic secondary schools, it was taught for a few hours a week. This instruction worked well, since, from the outset, it was built on our Latin competence. I was unlucky with English. I took private lessons, but after a few weeks the teacher got engaged and stopped teaching. Ultimately, I taught myself English, initially only enough to be able to understand the mathematical treatises, and more extensively later, in Palestine/Land of Israel. The result is embarrassing. My English pronunciation is poor. The style of the academic books I wrote in English is considered by some readers as "more Latin than English," which, however, did not seem to diminish their sales. Luckily, almost all of the talks and lectures I gave in English were in the United States, where people are very tolerant of poor pronunciation.

My experience with physical education was somewhat unusual. I felt a particular aversion to sports, and my father viewed anything that was not intellectual as a waste of time. With the help of a physician, I managed to be excused from physical education in the upper grades of secondary school. I would certainly have remained a home-staying narrow intellectual for my entire life had I not discovered, at the age of 15, my love of nature, which got me into hiking and later mountain climbing. In any case, in my childhood, I never dreamed that I would enjoy regular physical activity in the seventh and eighth decades of my life, nor that I would be awarded the gold medal for swimming in the state athletics examination and first prize at Hebrew University. Swimming was the only sport we were encouraged to do from an early age. After I turned 13, long-distance swimming became one of my goals. There were magnificent opportunities for swimming at the outstanding municipal swimming pools, and even more in the beautiful Upper-Bavarian lakes.

I do not have very much to write about the other school subjects. The assistant mathematics teacher, Herr Sondermayer, a genuine, crude Bavarian from Rosenheim, taught us basic mathematics, and later, physics. In terms of teaching skills, he was far superior to the main math teacher, and in the two years he taught us physics, I learned more in purely fundamental terms, than I did in Röntgen's class at the university.

The state-required two-hour weekly "religious instruction" was given in the senior classes by the main congregation's rabbi, Dr. Cosman Werner. Like almost all his colleagues, he was anti-Zionist, stressing that Judaism only had a religious character, and did not presume affiliation to any nation or race. Of course I categorically contested his view, not with ideological arguments, but with references to the Jewish Scripture. For me, this religious instruction was a farce, as confirmed by my having to write a composition for the *Abitur* examination on "Evil in the World in Light of the Bible."

The written *Abitur* examination took place on four consecutive days, without the now-common breaks to accommodate the students. There were six subjects: Latin, religion, German, Greek, French, and mathematics, with one question in physics. I left the mathematics examination before the time was up, after answering all the questions and commenting that the one in "mathematical geography" contained an unnecessary piece of data. The subject of the essay was also interesting, since it was typically "German": "To what extent do aspects of classical antiquity surface in the second heyday[7] of German literature?" In view of my great interest in Greek and Roman history, this was simple.

Although my *Abitur* certificate in July 1909 showed grades of "very good" in every subject, with special mention of the quality of my German essay and the scientific paper I had already published, I never capitalized on those successful examination results. In my experience, as confirmed by Justus Liebig and Albert Einstein, who were "poor students" but later became outstanding scientists, *Abitur* grades had no particular significance for later life.

However, I was extremely "ambitious," a vice that had good results. In my case, the "normal" adolescent drive to use up as much of their energy as possible was supplemented, and even surpassed, by my personal desire to please my parents. I especially wanted to get recognition and praise from my father, both in my regular schooling and in Jewish studies. I would have been ashamed not only in the eyes of

[7]This refers to the Classical-Romantic period.**

my teachers and myself, but even more so, my father, if I had come home with anything but the best grades.

In my childhood I understandably wanted to be successful in school. This was even reflected in my prayers. Since I did not find any suitable requests in the Eighteen Benedictions (*Amidah*), I oddly focused on the ninth blessing for "prosperous years,"[8] which actually refers to the rain and dew needed for agriculture in the Land of Israel. A conversation I had with my brother, sometime around 1905, is also typical. My brother and I assumed that emotion, reason, and will were three different types of intellectual activity. Not surprisingly, we both considered "emotion" to have the lowest value. I reinforced this view by attempting to "prove" that, for individual development, human will should be considered a higher and more essential value than reason. Indeed, my youth was characterized by the cultivation of will, and I will not presume to know whether that was good or bad. I referred to chapter 16, verse 1 of Proverbs (*Mishlei*),[9] which apparently contradicts the general assumption that, in prayer, one can voice expression, but only higher inspiration

---

[8]מברך השנים The prayer that blesses the years: "Bless on our behalf, O Lord our God, this year and all its kinds of crops for the best, and give dew and rain for a blessing on the face of the earth, and satisfy us from Your bounty, and bless our year like the best years. Blessed are You, God, Who blesses the years. . . ." ; adapted from *The Complete ArtScroll Siddur*. A new translation by Rabbi Nosson Scherman, co-edited by Rabbi Meir Zlotowitz (Brooklyn, NY: Mesorah Publications, 2007), 105.**

[9]The original quotation in Proverbs 16:1 is: לְאָדָם מַעַרְכֵי לֵב; וּמֵיְהוָה, מַעֲנֵה לָשׁוֹן,

"The reflections of the heart belong to man, but the tongue in which to express it is from the LORD." [The original text is obscure and can be interpreted in various ways. This seems to be Fraenkel's understanding of the verse]. The prayer to which Fraenkel refers is אוחילה לאל אחלה פניו, which appears in the *musaf* prayer on *Rosh Hashanah*, the Jewish New Year, and on *Yom Kippur*, the Day of Atonement.

אוֹחִילָה לָאֵל אֲחַלֶּה פָנָיו,
אֶשְׁאֲלָה מִמֶּנּוּ מַעֲנֵה לָשׁוֹן.
אֲשֶׁר בִּקְהַל עָם אָשִׁירָה עֻזּוֹ,
אַבִּיעָה רְנָנוֹת בְּעַד מִפְעָלָיו.
לְאָדָם מַעַרְכֵי לֵב, וּמֵה' מַעֲנֵה .לָשׁוֹן.
ה' שְׂפָתַי תִּפְתָּח, וּפִי יַגִּיד תְּהִלָּתֶךָ.

I shall put my hope in God, I shall beseech His presence,
I shall request of Him eloquent speech.
So that I can sing of His strength in the people's congregation,
That I can express glad songs for the sake of His human creations.
It is for man to arrange his feelings,
But eloquent speech is a Godly gift.
My Lord, open my lips, that my mouth may declare Your praise.

Translation from *The Complete Artscroll Machzor. Rosh Hashana*. A new translation by Rabbi Nosson. Scherman, co-edited by Rabbi Meir Zlotowitz and Rabbi Avie Gold (Brooklyn, NY: Mesorah Publications, 1997), 505. Fraenkel is claiming that in the conventional view of prayer, pertaining to will and aspiration, the content comes from God, and the mode of expression from man. By contrast, Fraenkel's view is that the ability to express oneself, and the voice, i.e., the means of the prayer, are given by God, but human will determines whether they are used to raise prayers to a meaningful level. Thus, while the means come from God, providing content and essence is the role of human beings. He found support for his interpretation in the verse from Proverbs in the אוחילה לאל prayer.**

fills it with substance. However, according to the interpretation in the mahzor (prayer book) for the High Holidays, speech or voice is a gift of God, but human will fills it with the devotion and the profound contemplation of the heart, which exalt it into a worthy prayer.

Visual memory is a good aid for students and teachers. In my mind's eye, I can still clearly see the pages with certain rules of Latin grammar or Homerian hexameter, and, in particular, sentences from the Roedelheim prayer book, the German translation of the *Chumash*, various chapters of the Talmud, and other sources. This phenomenon is well known among Talmudists. In some cases, I can still imagine myself "reading" as it were what was printed there in those texts, although I have not actually seen them for decades. However, such skills can only be developed by always using the same versions of the grammar book, siddur,[10] and *Chumash*. It definitely makes sense to promote visual memory in this way, and is certainly appropriate not only for those trying to develop eidetic imagery, such as Erich Rudolf Jaensch and others.

Already as a child I had a very mediocre memory. I remember past events inconsistently, as most people do. Thinking back to when I was 30, or even 40, I can generally remember most details precisely, even now, although in writing these memories I rely as much as possible on written documents. However, what I experienced later is becoming increasingly muddled. Fortunately, memory is not very important in mathematics, in which one thing always logically leads to the next. My memory would hardly have been sufficient to become a historian, biologist, doctor, or Talmudist. With one exception: Although I am definitely not musically inclined, I have a good "musical memory." For instance, I can easily remember Biblical verses with the cantillation marks (or *trop* or *ta'amei hamikra*), as well as verses in hexameter, especially in Greek, which sounds much more melodic to me than Latin or German. Thus, I know large portions of the Bible by heart, including virtually all the Pentateuch and most of the Latter Prophets, as well as parts of Homer's *Odyssey*. However, despite having memorized the three אמ"ת[11] books (Job, Proverbs, and Psalms) by repeated loud recitation, between the ages of 14 and 17, I forgot most of them rather quickly, because their equivalent musical cantillation is not known. Knowing text by heart, rather than simply knowing Biblical and Talmudic literature, helped me after immigrating to the Land of Israel/Palestine in 1929. It enabled me to give lectures in Hebrew immediately, with little preparation. I was far more familiar with ancient Hebrew literature and, to some extent, with Aramaic, through the Onkelos translation of the Pentateuch,[12] than I was with German literature, of which I had memorized very little.

[10]The siddur is the prayer book for every day and the Sabbath, not including services for the holidays.**

[11]This is an abbreviated form for איוב, משלי, תהילים **

[12]The Onkelos translation is the main translation of the Torah into Aramaic. This translation is attributed to Onkelos, a convert to Judaism around 35–120 AD.**

As already mentioned, around my fifth birthday I began studying the Pentateuch based on Rudolf Fuchs' translation. Herr Rotter gave me four hours of instruction a week. We started at the end of Exodus. In the Fuchs translation, the "suggestive" sections, for example, the stories of Lot and his daughters (Gen. 19:31–36), of Abimelech and Sarah (Gen. 20), of Joseph and Potiphar's wife (Gen. 39:7–20), and large sections of Leviticus (e.g. Lev. 18:6–23; 20:10–21), are rendered in small print without vocalization and translation. I did not learn those sections until after my bar mitzvah, by reciting the weekly portion aloud with the cantillation marks and the Aramaic translations. After we finished the Pentateuch, around my ninth birthday, he moved on to the Gemara, starting with the *Seder Mo'ed* tractates.

Thus, I largely learned classical Hebrew and acquired a sound knowledge of Targum-Aramaic in a playful fashion. I sometimes made active use of this, for example, to write letters in Hebrew. This was not done by translating German texts into Hebrew; rather, they were easy for me, since I knew the literary language. Only occasionally did I have to refer to the German-Hebrew index at the end of Gesenius's concise Hebrew lexicon,[13] as I had no access to a German-Hebrew dictionary.

In addition to the Gemara lessons I was given at home by Herr Rotter, from about the age of ten until the fall of 1910, and occasionally afterward, I had the good fortune to have regular private lessons in Gemara and the commentaries with Dr. Heinrich Ehrentreu, the great Talmudic expert and philologist. I particularly enjoyed the Tractate Sanhedrin, which demonstrated my teacher's unusual knowledge: for example, his grasp of the historical, astronomical, and climatological foundations of the Jewish calendar in the first chapter; and his philological and historical knowledge exhibited in the last aggadic chapter. Most members of Ohel Jakob, many of whom attended Ehrentreu's twice weekly adult education class, not so much out of interest but so as to satisfy a religious obligation, were hardly aware of his high standard of learning. Indeed, most of them did not realize how skilled Ehrentreu was until he was offered the rectorship of the Berlin Rabbinical Seminary and the Talmud chair after David Hoffmann died. Not until I studied elsewhere, starting in 1910, and took Talmud classes with other rabbis, did I realize the great difference in learning levels, which was sometimes like the difference between Hungarian and German styles of learning.[14] Almost all the truly significant Talmud scholars in Germany at that time were "Hungarians." However, some highly gifted non-local students in Munich, such as Fritz Gumpertz and Gerhard (Gershom) Scholem (both of whom have been living in Jerusalem for a long time), recognized Ehrentreu's greatness. To this day I consider it an honor to have received the title of *chaver*[15] from him.

---

[13] *Hebräisches und Aramäisches Handwörterbuch über das Alte Testament*; English: *Wilhelm Gesenius's Hebrew Lexicon of the Old Testament, including the Biblical Chaldee*, trans. Edward Robinson (1836; 5th edition, with corrections and additions, 1854).**

[14] Regarding the different styles of learning: see chap. 3, pp. 90–91, and chap. 5, p. 156.**

[15] The *chaver* title is conferred by a community rabbi on a community member for special service to the community.**

For my bar mitzvah in February 1904, I prepared the reading of the weekly Torah portion (תצוה, *Tetzaveh*[16]) and two speeches (*derashot*). I did not need any lessons for the former. Because I heard my father reciting the Shema with cantillation every morning, and on holidays and Sabbaths during the school vacations I heard the Torah reading at services, I quickly grasped the musical equivalent of the printed cantillation marks and was able to teach myself my *parashah,* the entire weekly portion. Herr Rotter and Rabbi Ehrentreu taught me the two *derashot*, to which I also made an original contribution. The first was from the Beitzah Tractate, and the second, for family reasons, was about a subject B. H. Auerbach, my great grandfather, dealt with in his main work: "נשים פטורות ממצוות עשה שהזמן גרמא, אך חייבות ב... "מפני שהן היו באותו הנס."[17] It was important for me to practice "*leining*" (reading from the Torah in the synagogue) in order to increase my command of the biblical text and language. In 1906 I had a chance to do that at the Sabbath afternoon "youth services," which I, along with others, organized for children who went to school in the morning. I also practiced "*leining*" during summer vacation, when no *baal koreh* (*leiner*) was available. Much later, in the Land of Israel/ Palestine, I was asked to be the *baal koreh*, especially in Hasidic synagogues.

More and more university students contributed to making our Friday evenings enjoyable. Unusual among Munich's religious circles, we invited them to join us for the Sabbath meal during the winter months. Discussions with the students provided me with opportunities to speak extemporaneously on Jewish and general themes when I was only 15. Among the guests I met before leaving Munich in the fall were the engineer Simon Gumpertz, who was Oskar von Miller's assistant; Max Michael and Fritz Gumpertz, who were friends; Dr. Nathan Weil; and Hedl, who later married Dr. Maurice Ettinghausen. Particularly notable was Selina Prins[18] from Amsterdam (1908), who later married David Sassoon. I had not yet met her younger sister, who was almost nine years her junior and would eventually become my wife. Selina lived in Munich with a long-established Catholic family, business friends of her father. As an Orthodox Jewish woman, she was given her own kitchen in their house so she could observe the strict Jewish dietary kashrut laws.

A particularly remarkable visitor to our parents' home in 1909 or 1910 was Dr. Moses Auerbach, son of Rabbi Selig Auerbach, who later moved from Munich, via Trieste, to Palestine/Land of Israel, where he served as director of the School of the Free Association in the *yishuv* in Petach Tikvah from 1910 to 1917. My parents' guest several times in later years, his accounts were among my earliest impressions

---

[16]The *Tetzaveh* portion is from Exodus 27:20 to Exodus 30:10.**

[17]This refers to an exception to the general rule that women are exempt from time-limited mitzvot (commandments). The exemption pertains to mitzvot relating to historical events in which women were involved. Note that women are obligated to follow regular non-time-limited mitzvot. The text says: "Women are exempt from time-limited precepts (mitzvot), but are obligated to [observe some others, pertaining to lighting *Hanukkah* candles, reading the Scroll of Esther on Purim, and drinking four cups of wine during the Passover *Seder*] because they too had been saved by those miracles."**

[18]See the Prins family tree.**

of the Land of Israel/Palestine. In 1935 Auerbach returned to the Land of Israel/ Palestine as director of a Mizrachi school.

As in many families in our circle, the holidays, in particular, Passover (*Pesach*), the New Year (*Rosh Hashanah*), and the Day of Atonement (*Yom Kippur*), were the key events of the year. I especially liked the atmosphere on the day before Passover (*Erev Pesach*), when, as the first-born, I fasted until midday.[19] Grandmother Fraenkel organized the baking of matzoh for the entire Fraenkel family on a day she reserved just for us at the bakery. I took an active part in this. I must have been 11 or 12 years old when I wrote a "play" in verse for Erev *Pesach*, my sole attempt at "poetry." With two "characters," "Workday" and "Yom Tov" (literally: holiday, in this case, Passover), it included a short scene in the middle when Yom Tov expelled Workday. It started with a monologue by Workday, followed by a longer one by Yom Tov, each recounting their characteristic traits. Of course, my brother played Workday, and I was Yom Tov. My brother did not resist my dictatorship.

On the High Holidays, my brother and I accompanied my father to the rather distant synagogue at 5:30 in the morning, often in driving snow. On the two days of the Jewish New Year, we remained there until the end of the service at about 11:30 AM; on *Yom Kippur*, starting at the age of 12, I fasted and stayed until the evening. These days made a deep mystical impression on me from early on. I was not so impressed by the main cantor, my beloved teacher Herr Rotter, as I was by Rafael Landauer, a lay cantor who had moved to Munich from Hürben near Krumbach in Swabia, with his brothers Isidor and Abraham. I can still hear his angelic voice and can recall all his melodies with the texts, both recitative and singing, in the *shacharit* (morning service). I was rarely so deeply moved by a cantor.

Since early childhood I wrote letters to my father for the New Year with greetings in Hebrew. In a long letter from the fall of 1901, which I still have, I see to my surprise that there is a letter attached to my mother, written in Jewish-German [i.e., Western Yiddish], in the Hebrew alphabet, which she was presumably able to read without any difficulty.

I experienced grave sickness and death for the first time after my 14th birthday. In March 1905, my sister Paula, who was not yet four years old, came down with double pneumonia and for ten days hovered between life and death. My parents worried day and night. My father and I read *tehillim* (psalms) every day. My brother, who was preparing for his bar mitzvah, was sick at the same time. Only later did I realize the incredible emotional tension my parents must have gone through. On top of everything, Grandfather Neuburger was suffering with a serious ailment and died in the summer of 1905. At this impressionable age I attended my first funeral. In late 1907, Grandfather Fraenkel died, leaving a tangible gap in my life.

---

[19]In commemoration of the last of the ten plagues, God's killing of the Egyptian first-born, it is customary for Jewish first-born to fast on the day preceding Passover. As a child, Fraenkel fasted only half a day.**

The birth of my sisters Else (in July 1896) and Paula (in May 1901) did not impress us older brothers very much. For a long time, like all boys at that time, we were totally convinced of our superiority over the girls. Things were different when our youngest sister Dora was born (in late 1904), since only a short time earlier, as incredible as it sounds, and seemed to me later, our father had explained to us a mother's role in birth. I essentially received my sex education only from the Bible and the Talmud. Even at age 16, I did not understand the reason for a scene between a teacher and a student who had made an inappropriate comment to the boy sitting next to him when in a Homer class we got to the hexameter: τερπέσθην φιλότητι, παρ' ἀλλήλοισι μένοντες.[20]

The Fraenkel siblings: from left to right, Abraham Adolf, Dora Dvora (on his lap), Joël Eugen, Tirza Paula, and Lea Else

Of course, when we studied the Pentateuch between the ages of about five and seven, there were enough "suggestive" sections for us to have learned a thing or two! Our attention was, however, not directed to those passages. And in any case, most of them remained untranslated in our book. Around 1950, my attention was

[20]This is part of Homer's description of Odysseus making love to the beautiful nymph Calypso just before parting from her (quotation set in italics): "So he spoke, and the sun set and darkness came on. And the two went into the innermost recess of the hollow cave, *and took their joy of love, abiding each by the other's side.*" (Homer, *The Odyssey*, trans. A. T. Murray, 2 vols. (Cambridge, MA: Harvard University Press; London: William Heinemann, 1919), book 5, lines 225–227.**

unexpectedly drawn to this issue. As chair of Hebrew University's Secondary School Commission, I was asked by the administration of the Arabic secondary school in Nazareth to visit their classes and make suggestions about their curriculum. A short time earlier, the Hebrew Bible had been introduced into the upper classes' curriculum; the teacher complained, for example, that in Genesis, chapter 18, which was part of the syllabus, menstruation was mentioned in verse 11 (as well as in 31:35). The students would giggle uncontrollably at these references. At first this upset me since I knew that no one objected to these and other far more suggestive passages in the Jewish schools. But the difference is easy to explain: Jewish schoolchildren learn these passages when they are six or seven years old, but the Arab students were 15 or 16!

Among the things that profoundly influenced my development were my religious background, the special experience of the Day of Atonement, Jewish study, and, after I turned 14, my interest in mathematics. Later, I discovered something else that inspired me: the Alpine landscape.

As previously mentioned, my physical education had largely been neglected from early on. My father felt that intellectual study should take priority and I myself had an aversion to exercise, considering it a waste of time that could be better used for study. The only physical activities I liked were the occasional games we played in our spacious yard and swimming. I also worked in our modest "garden" from the ages of nine until about 13. My grandfather, who owned a house and farm, set aside a small area for my brother and me to plant, and gave us some money to buy soil and plants. We planted mostly pansies, but some other flowers and vegetables too. In return for taking care of these undemanding flowers, we made a small profit each Friday by selling Sabbath bouquets to our mother and two grandmothers, who, depending on their generosity, paid from five to ten pfennigs for each. At the end of the season, I would write an annual report, listing all the receipts and expenditures.

Our walks and outings, for several weeks during our annual summer retreats until 1906, did not provide any real exercise. Since the summer vacation was long enough to make it worth our while to leave the city, we would sometimes rent a small house in an Upper Bavarian village, where my mother would run the household with the help of one or two servants. Kosher meat was sent by train from Munich several times a week, and we boys picked it up from the station with a small wagon. My father, who did not think much of this modest country life, usually preferred to go to spa resorts, where there were Jewish restaurants and where the family sometimes joined him.

My love of nature was kindled by an insignificant outing. On Sunday afternoons, my brother and I would sometimes accompany our father on trips to the Bavarian Upland, where he inspected wool for small sheep farmers in the area. On one such occasion, which must have been before the end of the [nineteenth] century, while walking on one of the nearby hills, we came across an unexpected view of Starnberger Lake, about 15 miles south of Munich. Generally not all that spectacular, the view nevertheless made a deep and lasting impression on me.

In 1901, I took my first major hike, almost 19 miles from Siegsdorf via Mauthäusl to Reichenhall. Our two trips to the North Sea, however, I found utterly boring, except for a short daily swim.

My brother received a bicycle for his bar mitzvah in 1905, and, with some begging, I did too. We would take summer bike rides into the countryside, starting from Munich or Tutzing. These trips awakened in me an appreciation of the beauty of the Upper Bavarian landscape, which soon became a familiar and memorable "home" for us.

I gradually went from hiking on the plains to easy mountain hikes. In the summer of 1907, I went with two friends on a week-long hiking trip to Innsbruck, which I remember as the most beautiful city in Europe. Salzburg pales in comparison. We started in Füssen and, via Fernpass, Kaiserhaus, and Rotwand, made our way to the Schliersee Lake. I planned the entire tour down to the last detail, including the hiking times, train schedules, overnight accommodations, provisions, and precise estimate of the costs in German and Austrian currency. I was so pedantic in organizing our hiking and resting times that my companions were sometimes justifiably annoyed.

In the summer of 1908, my life changed. I was no longer a pale stay-at-home student stooped over books, but instead became an "athlete" and mountain climber. I started doing exercises at the age of 50, and remained an impassioned hiker even at 70 and beyond. This transition was triggered not by an inclination or aptitude for sports, but rather by a passionate love of the Alpine landscape. Among my siblings, only my sister Paula shares this passion with me.

On the suggestion of our uncle Dr. August Feuchtwanger, my mother rented a very simple house on the Alte Landstrasse (Old Highway) between Bischofswiesen and Berchtesgaden for our summer vacation in 1908. Although the Berchtesgaden Alps are only moderate in height, under 10,000 ft, and therefore have very few glaciers, they are the most spectacular part of the Bavarian Alps, on a par with the Austrian Tyrol and the Swiss Alps. When the often unfavorable weather in the Northern Limestone Alps permitted, my brother and I took one- and several-day hiking trips in this border area between Bavaria and Salzburg. Only once were we accompanied by a guide. Otherwise, we used an excellent travel-handbook that focused specifically on this section of the Alps. We visited the Obersalzberg, which later became infamous,[21] and from there went mountain climbing. We virtually lived out of our knapsacks, without stopping anywhere for meals. We paid 80 hellers[22] a night at most to stay in alpine huts, which came to about 65 pfennigs per person. We would hike for 8–13 hours a day. Throughout my life it was vertical cliffs and ice walls, gravel deserts totally devoid of organic life, and ravines (called gorges in the western Alps) with powerful waterfalls that have impressed me much more than views of Grossglockner and Mont Blanc from 13,000-ft peaks. I always felt the greatest satisfaction when hiking alone. The feeling of being enveloped by rocky outcrops moves me much more than the "emotions at the peak" described in literary passages. Alone, I would instinctively be tempted to recite or sing sections from the Bible, such as Psalm 104, and the *piyyutim* (liturgical poems).

---

[21]The Obersalzberg, the Eagle's Nest, was Hitler's mountain retreat.**

[22]A heller was a German coin of low value worth half a pfennig.**

Unfortunately, on long climbs, even on comfortable trails, I always experienced painful breathing or heart problems, especially in the first two hours of uphill hiking. Although these difficulties diminished, they remained with me all my life. To keep going, I would force myself to do amateurish, yet effective, breathing exercises and maintain a slow tempo on the downhill stretch. The descents were very easy for me as long as they did not involve scrambling. On average, I would climb down in about a quarter to a third of the time that it took me to get up the slopes. I usually spent about three to four weeks a year in the high mountains, and over time, through persistent training, I succeeded in expanding the volume of my lungs and developing very strong leg muscles. While danger is not the "goal" of Alpine mountain climbing, it gives one an enjoyable "kick." Furthermore, for someone involved in intellectual work, the need to concentrate completely on climbing and avoid distractions is very relaxing. Letting my thoughts stray to mathematical problems could have cost me my life.

This training later earned me my reputation as an athlete, for example in my position as chairman of the Hebrew University's athletic commission and when I achieved the State of Israel Golden Sports Award. Only at an advanced age did I start doing other physical activity including exercises and yoga. In addition to lessons from some excellent female physical education teachers in Jerusalem, I was fortunate to get instruction from an Indian yogi, who visited Jerusalem around 1960.

The booklet entitled *Sport und Kultur* by Heinrich Steinitzer, published in 1910 in Munich,[23] gives a comprehensive, accurate definition of the term "sport." After correctly criticizing several dozen "definitions," he reaches the conclusion that it is not specific activities that make something a "sport," but the competition of physical strengths that occurs when performing them. He then applies this specifically to the culturally worthwhile practice of mountain climbing. The booklet, which also mentions the Jewish and Chinese stance toward athletics, created a great deal of protest among the tens of thousands of Munich Alpinists. As a member of the Munich section of the German and Austrian Alpine Association, from 1909 until 1933, I participated in some of those heated discussions.

At the northern edge of the Alps (Northern Limestone Alps), on the plateau where Munich and Berchtesgaden also lie, the weather conditions are particularly unfavorable and difficult to predict. When I had just started cycling in 1905, and especially when I went on mountain routes, I would try to figure out an accurate forecast. I read a lot of the popular literature on this, observed air pressure and cloud structure, and gathered my own experience about local weather developments, using a barometer and cloud atlas. I still have one of my notebooks, "Meteorological Notes III" (there must have been two earlier ones), with a record of the meteorological data for every day from May to September 1908 and from April to late May 1909, in Munich or Bischofswiesen. These included measurements of

[23]Heinrich Steinitzer, *Sport und Kultur* (Munich: Deutsche Alpenzeitung publishers, 1910).

atmospheric pressure with a table I prepared to calculate its reduction to sea level, the quantity and type of clouds, and the precipitation. All those were recorded three to four times a day. I even recorded the moon phases, presumably to show that the moon does not influence the weather.

This turned me into a successful forecaster, not in a scientific sense, but based entirely on intuition. Later, in the First World War, I served as an army meteorologist for two years, and after 1930 I used these skills to take several brief trips to the Dead Sea, which is notorious for frequent and dangerous winter storms.

Fraenkel, on the left, with a guide in the Alps

After finishing secondary school, in the summer of 1909, I got permission to hike alone in the Alps and for the first time experience real mountaineering, sometimes with ropes. During the first few weeks, I went alone or with a guide, after crossing over the Ortler Alps (Cevedale) to the Bernina range, and then spent some unforgettable weeks in the Upper Engadin and the area around Pontresina. I went on some of the most beautiful hikes and alpine routes in the Bernina range and the Berchtesgaden Alps. Except for the really high-altitude routes, I usually hiked in the Bavarian Alps with my brother. I was not only the planner, but also the guide, perhaps a little too dictatorial, but willingly tolerated. In 1910, I had the pleasure of persuading my mother, who was already 43 years old, to take a mountain hike, for the first and last time in her life. She enjoyed it tremendously. It was not even all that difficult for her, since she was agile and light.

In August 1919, shortly before I got engaged, I went on a family trip with all my siblings to the Reiter Alps, near Berchtesgaden. It was a kind of farewell to my youth. Later I preferred the Swiss Alps. I took my last mountaineering trip with a guide and ropes, in the Bernese Oberland, in 1961, when I was over 70. Actually it was a bit too much for me.

It may be surprising that before I chose my future profession (before studying mathematics and publishing my first articles), the impetus for my choice of mathematics lay in my intensive Jewish studies. As bar mitzvah presents, I received many Hebrew books and works dealing with Jewish themes, and I expanded and improved this collection through lively exchange with friends. In the winter of 1904–1905, around my 14th birthday, during the choral singing in the synagogue, which I found rather boring, I would read the "*Tiferet Yisrael*" commentary to the second section of the Mishnah, *Seder Mo'ed*, on the Jewish holidays. Among the comprehensive commentaries, I found one with the misleading title "*Shevilei de'Rakiya*" (Orbits of the Heavenly Bodies). The author Israel Lipschitz correctly described the aim of this commentary being to teach people how to calculate the calendar. Although not entirely free of error and not very original, the tractate is written with great pedagogical skill. I was fascinated and would later expand on this subject, which was still very new to me at that time.

The career of my uncle by marriage, Alfred Loewy (1873–1935), a distinguished mathematician, is meaningful to me personally. It is also of interest because it illustrates the issue of Jews at German universities, which I will touch upon several times. Very talented, extremely diligent, and almost puritanical, he came from a merchant family from Rawicz (Posen, now Poznań, Poland). From 1891 to 1895, as a young man, he studied at universities in Breslau, Berlin, Munich, and Göttingen, and received his doctorate from the University of Munich. His advisor, Professor Ferdinand Lindemann at the University of Munich's School of Arts and Humanities, recommended him for a position to acquire his professorial qualification (*Habilitation*) to become an unsalaried adjunct lecturer (*Privatdozent*). Although the department's hiring practices were relatively liberal until the First World War, and Lindemann had considerable influence, Alfred Loewy was rejected, not as a Jew per se, but rather as a rare case of an Orthodox Jew in academia: Anyone who does not eat with the department staff for reasons of kashrut, they commented, should also not work with them. Also on Lindemann's recommendation, he was hired by Professor Jacob Lüroth at Freiburg University as an unsalaried adjunct professor in 1897. Though not a brilliant educator, and even less so a "light" author, as he pedantically sought to explain everything down to the last detail, he was in fact an excellent researcher and instructor of advanced students. However, when a professorship opened up, he was passed over as a Jew, even in the liberal "model state" of Baden, in favor of a less qualified candidate, and failed to find work at a foreign university. Only after the German Revolution, in 1919, did he finally receive his well-deserved promotion to full professor (*Ordinarius*) in Freiburg. Although he lost his eyesight in the 1920s, he continued his research and teaching until he was forced to retire when the Nazis took power. A short time later, he died in the aftermath of an operation.[24]

In October 1906, Alfred Loewy, a non-tenured associate professor (*Extraordinarius*) since 1902, married Aunt Therese, my mother's much younger sister. While my uncle could not help me in a professional sense, he did give me advice and ideas. He reinforced my interest in the Jewish calendar and drew my attention to the formula[25] for calculating the date of Passover in the Christian calendar, which Carl Friedrich Gauss, the "Princeps mathematicorum" (the Prince of Mathematicians), had published in 1802, though without any proof. To test my skills, Loewy suggested that I attempt to find a proof for this formula. Experts had been trying to do this for decades, but in vain. Actually, in 1896, Meyer Hamburger (1838–1903), a renowned mathematician at Berlin's Technical University, furnished an elegant proof. Unlike his fellow classmates in Posen, Lazarus Fuchs and Leo Königsberger, both of whom soon became full professors in mathematics, Hamburger rejected the convention in Prussia at that time of using baptism as a means of opening up more options in his academic career or even gaining access to higher teaching positions. Throughout his life, he earned his livelihood as a teacher at the boys' school of the Berlin Jewish Community.[26]

In 1907, I sent an analysis of the Gaussian formula to Loewy. Although not a full proof, it did reveal the main ideas on which the formula was based. Loewy was always a strict critic, but his response sufficiently encouraged me to take on new, relevant problems once I had studied Hamburger's proof. Consequently, in 1907, shortly after my 16th birthday, I completed my first independent research project: the development and proof of a formula for calculating the date of Passover in the Islamic calendar.[27] Of course, the part which refers to the Jewish calendar used the Gaussian method. I do not know if this paper would have ever been read if it had not aroused the interest of Professor Siegmund Günther, whom I mentioned earlier. He arranged for publication of my paper in 1908 in the *Zeitschrift für Mathematischen und Naturwissenschaftlichen Unterricht* (Journal for Mathematical and Natural Scientific Instruction), after I complied with the publisher's request to shorten it

[24]I briefly sketched his scientific significance in an article published in English in *Scripta Mathematica* 5, no. 1 (New York, January 1938): 17–22.

[25]Of course, all these formulas only have theoretical significance. In practice, data can be compared more rapidly by using tables.

[26]See the annual report of the Deutsche Mathematikervereinigung (German Mathematical Society), vol. 13 (1904): 40–53.

[27]This is a strictly lunar calendar without any connection to the solar year. Herein partly lay the difficulty of the problem, which despite numerous attempts had not been previously solved.

by one third.[28] I subsequently received numerous letters, on the one hand, from Jews living in Muslim countries, including Palestine/Land of Israel,[29] such as Elieser Grünhut in Jerusalem, and the secretary of the Grand Rabbin de la Turquie (Chief Rabbi of Turkey); and, on the other hand, from European mathematicians interested in calendars, who addressed me as Doctor or even Professor, which, of course, was flattering for a 16-year-old high school student.

Much more important was the research project I completed during the summer break before I started studying at university (in September/October 1909). It was a mathematically insightful paper on another calendar problem, first treated by Gauss, which, thanks to Loewy's referral, was soon published in one of the most prestigious international journals, the *Journal für die Reine und Angewandte Mathematik* (Journal of Pure and Applied Mathematics).[30] After wavering between mathematics and classical philology, that paper clinched the decision for my future profession in favor of mathematics.

My equally strong preference for classical philology and mathematics perhaps explains my aversion to the organizational structure of most schools and universities, which strictly separates philological history from the areas of mathematics and the natural sciences. Even if the "natural sciences" are understood to denote physics and chemistry, and do not include systematic biology, which is essentially different and given only a modest place in the curriculum, this bifurcation of students' talents is certainly unsatisfactory. There is a small class of high-school students who are not interested in historical description and archaeological finds, nor in carrying out experiments, but only in theoretical results. These students—in which quality balances out quantity—are fascinated by the laws in mathematics, theoretical physics, and chemistry, which they also find in ancient languages and comparative philology. Gauss, Carl Gustav Jacobi, and other outstanding mathematicians belong to this class, even if they were also interested in experimentation to some extent. Hermann Grassmann is a classic example. Disappointed because his mathematical genius was not recognized by his contemporaries, he turned to comparative linguistics, in which he made a great name for himself.

Thus, I decided to study mathematics as my major and attempt a career in academia, despite the limited advancement opportunities for Jews. My father supported my decision, but because I was so young, he wanted me to spend my first two semesters living at home. This meant studying at the highly esteemed University of Munich, the second largest university in Germany. My brother fulfilled my father's wish for a son to take over the family's renowned wool company, founded by my great grandfather. He joined the company after earning his doctorate in economics, as was common practice among Germany's upper

---

[28]A popular presentation of the results appeared as "Eine Formel zur Verwandlung jüdischer Daten in mohammedanische" (A Formula for Converting Jewish Dates into Muhammadan Dates) in the journal published by Markus Brann, *Monatsschrift für Geschichte und Wissenschaft des Judentums* 53 (1909): 736–743.

[29]The Land of Israel was under Ottoman rule at the time.**

[30]A. Fraenkel, "Die Berechnung des Osterfestes," *Journal für die reine und angewandte Mathematik* (commonly referred to as *Crelle's Journal*), no. 138 (1910): 133–146.

middle class Jews. I chose physics and philosophy as my minor subjects; physics, despite my limited interest in experimentation, because it is objectively and traditionally linked with mathematics; and philosophy because I had been interested in philosophical questions since high school, and also because of its relationship to questions of religion. At that time, I had no idea that I would ultimately drop philosophy as an examination subject, or that, after more than a decade, I would move directly from the foundations of mathematics to logic and would later, without any real justification, be considered a logician.

The year I spent studying in Munich was more like the conclusion of my youth within my family circle, rather than a case of independent integration into academia. At that time, the atmosphere at the University of Munich was largely determined by the students from northern Germany who were attracted by the beer-drinking, the carnival, the Bohemians of Munich's Schwabing district with their artistic ambitions, the light-hearted southern way of life, and sometimes also the Alps, which were only 40–60 miles away. Of course, not only did this atmosphere not draw me in, but also, as an Orthodox boy sheltered by his family, it remained totally foreign to me, except for the mountains. Already in my final year of high school, I only reluctantly, as an unwilling outsider, participated in the school's officially organized "beer pubs," whose crude entertainment was accompanied by the singing of truly German student songs.

Even my hopes of finding scholarly stimulation in the first year of my studies at a university with world-renowned names on the faculty turned out to be elusive. The reasons for this say less about me than about the history of the University of Munich's School of Arts and Humanities.

Of the 20 German universities, Munich was among the few with three full professors in mathematics. Other universities generally had only two, and some had only one. The teaching staff also included an honorary professor, an associate professor in geometry, and three unsalaried adjunct professors.

Until the First World War, the University of Munich was nevertheless rather weak in mathematics, in both teaching and to some extent research, compared with other German universities, as a result of objective circumstances and personnel. The major bone of contention was the Bavarian examination for upper-level secondary school teaching candidates, which was the goal of most mathematics and physics students. The Prussian requirements, initiated as early as the mid-nineteenth century by Peter Gustav Lejeune Dirichlet and Jacobi, enabled and even encouraged candidates to get scientific training. By contrast, the examination questions in Bavaria were mainly based on the secondary school curricula and standard introductory lectures. Thus, unlike the situation in Prussia, the final examination essentially consisted of written exercises. Accordingly, the higher-level mathematics lecture classes had only a small number of students, since, unlike for the other subjects, most of them were Bavarians. This was even more so in the University of Würzburg's Natural Science Department. However, the third and smallest Bavarian university, at Erlangen, which was well known mainly for the Protestant theologians there, also had an international reputation in mathematics, thanks to the distinguished researchers Paul Gordan and Max Noether. The latter was Jewish, and the father of the most outstanding female mathematician, Emmy Noether.

Incidentally, the full professors in the University of Munich's Mathematics Department, Aurel Voss, Ferdinand Lindemann, and Alfred Pringsheim, were all non-Bavarians, as was most of the teaching faculty. They were all relatively advanced in age and, except for Pringsheim, no longer very productive. Since Voss, a sophisticated scholar, who was 65 at the time, was very sick and often had to interrupt his lectures, I discuss only Lindemann and Pringsheim in this book.

F. Lindemann

Ferdinand Lindemann, originally from Hannover, was considered somewhat peculiar, which is a common trait among mathematicians. He was by no means one of the great scientists of his time, but earned world fame by finally solving one of the oldest and best-known mathematical problems: squaring the circle. The problem goes back to the fourth century BCE. Based on the decisive work of Charles Hermite of France, who, with great acumen, proved the transcendence of the number e (the base of natural logarithms) in 1873, Lindemann proved in 1882 that π (the ratio of the circumference of a circle to its diameter) is also a transcendental, i.e., not an algebraic number, and thus could not be constructed with a straightedge and compass, as was already known.[31] Owing to this discovery, he was appointed to a professorial position at the University of Munich in 1893, when he was only 41 years old. As it later turned out, this appointment was inappropriate, at least with regard to his research and teaching activities. Lindemann's later treatises had only limited significance. He became known for his repeated, and without exception faulty, attempts to prove "Fermat's last theorem," published in articles and as a monograph. This theorem, which concerns the Diophantine equation, $x^n + y^n = z^n$, was conjectured around the mid-seventeenth century by the brilliant French number theorist Pierre de Fermat, who was a lawyer by profession. He claimed to have discovered "a truly wondrous proof" (*demonstrationem sane mirabilem*) for this. The theorem has occupied numerous great mathematicians for more than three centuries, to the present day. Although particularly in the nineteenth and twentieth centuries it has been proven for many, or even an infinite number of values of the exponent n, by means of extremely profound methods not even dreamed of by Fermat, the problem has remained unresolved until today. We do not even know if the theorem is *true*, in general, although this is highly likely, even though Fermat's "proof" was certainly faulty.[32]

In the early twentieth century, in addition to the scientific interest in this theorem, a "financial" incentive was introduced. In his will, Dr. Paul Wolfskehl,[33] a Jewish amateur mathematician from Darmstadt, provided a prize of

---

[31] Transcendental numbers and squaring the circle:

As the Pythagorean school already discovered in the sixth century BCE and initially treated as an esoteric secret, there are "irrational" numbers, which cannot be expressed as the ratio of integers $\frac{m}{n}$. The simplest such number is $\sqrt{2}$, the length of the diagonal of a square with a side length of 1.

Since $\sqrt{2}$ is a root of the equation $x^2–2 = 0$, it seemed reasonable to assume that all irrational numbers could be expressed as roots of "algebraic equations" of the form:

$$a_0x^n + a_1x^{n-1} + \ldots + a_{n-2}x^2 + a_{n-1}x + a_n = 0$$

with $a_0, \ldots a_n$ integers. All such roots are called "algebraic numbers."

In 1851, this assumption was shown to be false: thus, there are also non-algebraic numbers, which are called "transcendental." The best-known transcendental number is π, and Lindemann proved its transcendence in 1882.

Only quantities that solve linear and quadratic equations ($n = 1$ or $n = 2$) can be constructed by using a straightedge and a compass. These are special and the simplest kind of algebraic numbers. Because a circle with a radius r has a circumference of $2\pi r$ and an area of πr^2, it is impossible, for example, even with far more complex instruments than a straightedge and a compass, to square the circle, that is, to construct a square with the same area as a circle of a given radius.

[32] Fermat's theorem was finally proven in 1995, after more than 350 years, by Andrew Wiles of Princeton and Oxford.**

[33] In series 3, vol. 4 of the *Archiv der Mathematik und Physik*, around 1902, Wolfskehl set a calendar exercise, namely, to determine the proof of one of the formulas he gave for calculating the date of the Gregorian Easter. For many years, the problem remained unsolved until I became aware of it as a young student in 1910, and solved it immediately. The proof I proposed was published in the same journal in 1911, in series 3, vol. 17. The solution was long in coming not because the

100,000 gold marks (about 24,000 gold dollars) to whomever could prove or disprove Fermat's last theorem. In 1907, the prize money was transferred to the Royal Society of Science and Humanities in Göttingen. As a result, hundreds of so-called proofs were sent to Göttingen by people of all social classes: from merchants and craftsmen without previous training to mathematics professors. However, the number of proofs submitted for review, and the inconvenience associated with reviewing all of them, declined rapidly when the prize money became virtually worthless due to the inflation of German currency from 1918 to 1923. Nevertheless, about a dozen proof attempts were sent to me in Jerusalem to review, even after 1929.

Lindemann's failure as an academic teacher was somewhat peculiar. From fall 1909 to summer 1910, I attended his main and fundamental introductory lecture class on differential and integral calculus, a required course for all first-year mathematics and physics students. Despite the elementary character of the subject matter, it often happened, as his students learned to expect, that he would "get stuck" and could not complete proofs, and would not fill in the gaps until the next session.

As an individual and a scientist, however, he was a real personality. His physical appearance was majestic, his face framed by an impressive beard. He was certainly far removed from the anti-Semitic currents, which did not fully unfold at the University of Munich until the 1920s, yet were felt earlier. It says something about his character that, as late as 1936, he sent greetings to me in Jerusalem by way of another Jewish emigrant. In particular, he provided great service to the university administration and thus enjoyed high esteem among his colleagues, even after he became a professor emeritus in 1923. He was granted noble status, and given the rare title of Privy Councilor (*Geheimrat*). He died at the age of 87, in 1939.

The University of Munich's Mathematics Department did not always have a less than satisfactory ranking. Around 1925, for example, the three full professorships were held by the outstanding Constantin Carathéodory, Oskar Perron, and Heinrich Tietze.

More interesting, especially for the contemporary reader, is Alfred Pringsheim, who was the third full professor while I was studying there. He was a Jew who had legally abandoned the Jewish religion and was regarded as "religiously unaffiliated." His Jewish origin explains why he was appointed to a professorial chair so late, not until 1901. He was born in 1850 into a family of scholars, with many branches, in Ohlau (present-day Oława), Silesia. He was a distinguished theorist of functions of a complex variable. Also a patron of the arts and a talented musician, he was friends with Richard Wagner and Franz Lenbach. With a sizable fortune and great knowledge, he amassed one of the most valuable art collections in Munich. Among the generally dry and humorless German mathematicians, he stood out, not only for his expertise, but also because he was just about the only one with a sense of humor. Thus, in spite of his Jewish appearance, he was often asked to deliver the humorous speech that was typically given at the social banquets of the annual conventions of the German Mathematical Society. He became well known,

---

problem was particularly difficult, but apparently because no mathematician who was also knowledgeable about calendars had shown an interest in it.

in some people's eyes even notorious, for the speech he gave in 1924, in which he lambasted two eminent colleagues: Professor Ludwig Bieberbach of Berlin University, who in 1934 propounded the "theory" that "German mathematics" was outstanding in contrast to inferior "Jewish mathematics"; and his own former student, Professor Edmund Landau of Göttingen University.

Bieberbach had published a textbook, containing a rather arrogant preface, on the theory of complex functions, which was Pringsheim's domain. Pringsheim responded to this in his speech, with a smile, saying that surely his colleagues knew that English schools referred to "elementary geometry" as "Euclidean." In a similar vein, he said he had received word from England that they were now considering abandoning the term "theory of functions" and instead simply refer to what they pronounced as "Byberbatch."

Landau, who published more than any of his contemporaries, wrote, as I did, a short article for the mathematics-physics section of the *Scripta Universitatis atque Bibliothecae Heirosolymitanarum* (Jerusalem), in which all the articles appeared in Hebrew translation as well as in their respective native European language. It included articles by such well known scientists as Harald Bohr, Jacques Hadamard, Albert Einstein, Tullio Levi-Civita, and Theodore von Kármán. This section of *Scripta* was edited by Einstein in name only, but de facto by Immanuel Velikovsky, who later became widely known for his sensational, odd books, particularly *Worlds in Collision* (1950). Alluding to Landau's publication, Pringsheim pointed out that Landau admitted that he could not quite manage his publications by only writing from left to right, as usual, so he started writing also from right to left, in Hebrew.

As a millionaire, Pringsheim could afford to lecture[34] about really interesting subjects, instead of giving introductory lectures for large numbers of students. For that reason, I did not take any classes with him in my first year, but in the summer semester of 1912, my last semester in Munich, I took his course on analytical number theory. A small, lively man, he was extremely impressive. His class, interesting but somewhat difficult, had only eight students, including one woman. After three weeks, it apparently got too hard for the woman, who stopped coming. At the beginning of the next class, Pringsheim rubbed his hands in glee, and with a gleam in his eyes, said, "Well, gentlemen, it looks like it's just us boys again."

In the 1920s, I often met Pringsheim when I visited Munich. His life ended in a very sad and unexpected way. From 1922 on, he was an emeritus professor and initially enjoyed his scientific and artistic *otium cum dignitate* ("leisure with dignity") to the fullest. He continued to publish scientific articles until he was 83. From 1933 on, he suffered greatly from humiliating attacks by the Nazis. He was forced to "sell" his splendid, artistically decorated villa ("Pringsheim Palace") on Arcis Strasse to the Nazi party because they planned to build the "Führerbau," offices for Hitler and his closest staff, on that site. After refusing to swear the oath to the Führer, which he was required to do as an emeritus professor, he was forced to retire. After years of negotiation, he was granted permission to transport and sell his valuable majolica (tin-glazed pottery made in Italy) collection in London, provided

[34]See explanation of lecture fees on pages 67–68.**

that he hand over most of the proceeds. The moderate sum he was allowed to keep permitted him, at the age of 90, to travel to Zurich with his wife, shortly after the war started, where they set up a modest household. He died in Zurich in 1941.

Pringsheim Palace

As is well known, Pringsheim's daughter Katia married the writer and later Nobel Prize laureate Thomas Mann in 1905. Incidentally, Katia's mother was the daughter of Ernst Dohm, satirist, journalist, and founder and publisher of the satirical Berlin weekly *Kladderadatsch*, and the writer and women's rights activist

Hedwig Dohm. Undisputedly, the marriage had a rather decisive influence on Mann's political views. Thomas Mann, who had developed nationalist leanings during the First World War, moved more and more to the left in the 1920s.[35] After the Nazis seized power, he not only left Germany but attacked its politics in very harsh terms. Even after the war, he adamantly turned down members of German literary circles who called upon him to return. My wife and I met with him and his wife in the 1920s, in Marburg and in Jerusalem. In late March 1930, after having undergone an operation at Jerusalem's German Hospital, he asked us to pay him a visit.

Finally, I would like to commemorate two excellent mathematicians who were unsalaried, adjunct professors (*Privatdozenten*) at the University of Munich during my first year there (1909–1910). The first is Friedrich Hartogs, born in 1874 in Brussels. He was Jewish, but he no longer had any connection to the Jewish community. As a result of his original research on the theory of functions, he became a full professor at the age of 53. We shared scientific interests and research in the area of set theory. Throughout his life, he was shy and even anxious, and as it turned out, this was not without good reason. He suffered terribly after the Nazis came to power, and, as far as I know, took his own life during the war. The second, Oskar Perron, was born in Bavaria in 1880. Thanks to his outstanding research, he made a career for himself rather quickly. In 1914 he became a full professor at Heidelberg University, and in 1922 he succeeded his teacher Pringsheim in Munich. Perron clearly disassociated himself from the Nazi regime both before and during the war, which certainly took courage and put him at risk. Immediately after the war he took advantage of his first opportunity—for example as a member of the Bavarian Academy of Sciences—to publicly denounce the injustice suffered by the Jews and half-Jews among his colleagues. He also maintained his contact with me until the war started. Then it became too risky to write letters to Palestine/Land of Israel, but he resumed the correspondence immediately after the war. We met again at the International Congress of Mathematicians in Amsterdam in 1954. He continued his research for a long time, well beyond his 80th birthday—a rare exception among mathematicians.

My friend Arthur Rosenthal did not complete his postdoctoral professorial qualification (*Habilitation*) at the University of Munich until 1912. He was born in 1887 into a respected Jewish family in Fürth, Bavaria. He too was an eminent mathematician, and in 1922, he succeeded Perron as professor, initially as a non-tenured associate professor, at the University of Heidelberg. Dismissed by the Nazis in 1935, he, like so many other mathematicians and physicists, soon

[35]See, for example, the brochure with a preface by Thomas Mann, *Kampf um München als Kulturzentrum* (The Struggle for Munich as a Cultural Center), Munich: R. Pflaum, 1926. It includes six speeches (by Thomas Mann, Heinrich Mann, Leo Weismantel, Willi Geiger, Walter Courvoisier, and Paul Renner) given in an overcrowded lecture hall in Munich. From Thomas Mann's speech: "Unfortunately it has well-nigh reached the point that anyone in Germany who displays any trace of saneness is thought to be a Jew and thus is done for." This publication spurred a correspondence between Mann and myself.

found an opportunity to teach in the United States. I met him in 1946 at the University of New Mexico, and in 1951 at Purdue University, when I lectured at those universities, not without his help. Later, we met again in the Bernese Oberland, Switzerland. His name is permanently linked with the theory of real functions. He died in 1959.

At this point, I would like to mention Hugo von Seeliger, one of the greatest astronomers of his time, whose work is still significant today. He received a professorship at the University of Munich in 1882, when he was only 33. I attended only an introductory lecture course he gave in 1909–1910, and was profoundly impressed by his scholarly nature. I had a personal relationship with him through my calendar publications, which, although purely mathematical, are connected to astronomy. He is also known as the definitive teacher of the astronomer Karl Schwarzschild.

In my view, Karl Schwarzschild should be regarded, besides the mathematician Hermann Minkowski, as the only truly brilliant Jew in the combined field of mathematics and physics in our time. Born in 1873, he came from an affluent, religious merchant family in Frankfurt am Main. Other members of this family were Emanuel Schwarzschild, one of the founders of the Israelite Religious Society (IRG), the secessionist Orthodox congregation that appointed Samson Raphael Hirsch as their first rabbi; and Henriette Schwarzschild, who later married Moritz Feuchtwanger of Munich, cofounder of the banking firm J. L. Feuchtwanger, which still exists today.[36]

Karl Schwarzschild

Schwarzschild was an astronomer, but by all means eclectic. In the period from 1890 to 1916, he published about 100 scientific works, covering pure and applied mathematics, theoretical physics, meteorology, precision mechanics and apparatus engineering, and, in particular, theoretical and observational astronomy and astrophysics. Already as a high school student, he published articles in the

[36]See footnote 10 in Chap. 1, page 4.**

Astronomische Nachrichten, the leading astronomy journal in Germany. He was introduced to theoretical research first and foremost by Hugo von Seeliger in Munich, who was also his advisor for his doctorate and, in 1899, for his postdoctoral professorial qualification. At the age of 28, he became a full professor and director of the observatory at the University of Göttingen, where he brought astronomy into close contact with related disciplines. In 1909, he was appointed director of the largest observatory in the world, the Astrophysical Institute in Potsdam.[37] Although he refused to be baptized, which was a prerequisite for the position, he was nevertheless appointed, and remained there until his early death.

In the final years of his life, he was concerned mostly with problems related to the theory of general relativity, discovered by Einstein not long before. According to Schwarzschild, his own findings made Einstein's theory "shine with increased purity." But his related research, some of which was written in the field, remained fragmentary. When the First World War began, he joined the military weather service and, later, an artillery unit responsible for calculating trajectories. However, he developed pemphigus, a malignant autoimmune skin disease, while on the Eastern front, and died on May 11, 1916. On the same day, his last significant work, "On the Quantum Theory," was published in the proceedings of the Prussian Academy of Sciences. Schwarzschild was anything but an otherworldly scholar. Young and spirited, he was enthusiastic about art and poetry, and was a great mountain climber. He lived a short but rich life.[38]

Carl Gustav Jacobi (1804–1851), James Joseph Sylvester (1814–1897), Gotthold Eisenstein (1823–1852), and Leopold Kronecker (1823–1891) were the greatest Jewish mathematicians of the nineteenth century.[39] However, in terms of originality and depth, in the nineteenth century, I think Hermann Minkowski, brother of the well-known diabetes researcher Oskar Minkowski, surpassed them all, as did Johann von Neumann in the twentieth century. Minkowski, born in 1864 in Aleksotas, Russia (today Lithuania), moved to Germany as a child, and later, in 1880, attended the University of Königsberg. In the first year of his studies, he heard about a Grand Prix sponsored by the Parisian Academy of Sciences to solve the problem of the number of representations of an integer as the sum of five squares. He submitted such a brilliant solution, going much beyond the question posed, that the academy overlooked all the regulations against the use of the German language and awarded the full prize to him, then a 17-year-old student. His favorite subject remained "number theory," but not only in the classical forms. He invented a totally new approach, the "geometry of numbers," with impressive results. He was also one of the first to recognize the revolutionary significance of set theory, proposed by Georg Cantor.

---

[37]It became the Leibniz Institute for Astrophysics in 2011.**

[38]For more details, see the obituaries by Arnold Sommerfeld, *Die Naturwissenschaften* 4 (1916): 453–457, and Otto Blumenthal, *Jahresbericht der Deutschen Mathematiker-Vereinigung* 26 (1918): 56–75.

[39]See A. A. Fraenkel, "Jewish Mathematics and Astronomy," *Scripta Mathematica* 25 (1960): 33–47.

Hermann Minkowski

Minkowski became a full professor at the University of Königsberg in 1895, as David Hilbert's successor. From 1896 to 1902, he worked at the Swiss Federal Institute of Technology (ETH) in Zurich, an institution that had already attracted many other first-class scholars from Germany. In 1902, a unique situation occurred in German academia: Hilbert, the undisputed genius among German mathematicians, demanded "out of the blue" that the Ministerial Director Friedrich Althoff, director of the Higher Education Department in the Prussian Ministry of Education and Culture, create a fourth tenured professorship for mathematics for Minkowski at the University of Göttingen. This professorship was indeed created. Incidentally, after Minkowski's death, Hilbert was again instrumental in getting another Jew, Edmund Landau, to fill the position. Later, in 1927, Landau became a professor at the Hebrew University for a short time, before returning to Göttingen.

Minkowski dedicated the last years of his life mostly to introducing revolutionizing ideas to the special theory of relativity, formulated by Einstein in 1904. On September 21, 1908, he gave a lecture on space and time in Cologne,[40] which began as follows: "Henceforth, space by itself, and time by itself, are doomed to fade away into mere shadows, and only a kind of union of the two will preserve an independent reality."[41] He never had the opportunity to further develop his program of a four-dimensional space-time continuum. Not yet 45 years old, he died from a ruptured appendix, on January 12, 1909. Hilbert's obituary, which appeared in the *Nachrichten* of the Göttingen Society of Sciences in 1909, offers a comprehensive picture of this scientist, describing him as "... a rare gift from heaven."

But to return to my own studies, in 1909–1910, I took Wilhelm Röntgen's weekly five-hour main lecture class on experimental physics, which was a

[40]At the 80th Assembly of German Natural Scientists and Physicians.**

[41]Hermann Minkowski, "Space and Time," in Hendrik A. Lorentz, Albert Einstein, Hermann Minkowski, and Hermann Weyl, *The Principle of Relativity: A Collection of Original Memoirs on the Special and General Theory of Relativity* (New York: Dover, 1952), 75–91, here: 75.**

mandatory course for students in mathematics and several other subjects. It is not up to me to determine the extent to which the discovery of x-rays was due to luck or to Röntgen's tenacity in his experiments with cathode rays in vacuum tubes. In any case, after the discovery in 1895 he examined the properties of the new rays with great diligence. In 1901, he received the first Nobel Prize for Physics. He was teaching at the University of Würzburg, in Bavaria, when he made his discovery, so it is not surprising that a few years later he was given a professorship at the University of Munich and showered with honors. He taught until he retired in 1920.

The lecture class I mentioned had a registration of about 400, mostly medical students, which was an unheard-of number at the time for a class in the School of Arts and Humanities. Today, it is easy to forget how small the number of students in the exact sciences was, almost everywhere until the First World War, and even until the Second World War. The number of students at German universities was not only a source of moral gratification or disappointment to the teachers, it was also primarily significant from a material standpoint. Strange though it might sound to western Europeans or Americans, students had to pay a considerable sum for every lecture, except for seminars ("*privatissime et gratis*") and the few lectures offered free of charge (usually "*publica*"). This so-called lecture fee, which, until the First World War, was four to five marks per hour per week, went to the instructor. Therefore, aside from special regulations of the government ministry for particularly deserving scholars, the professors in such areas as oriental studies and astronomy had to make do on their salaries alone. In classical philology, modern languages, mathematics, and theological subjects, however, they could significantly supplement their fixed salaries. Indeed, in disciplines such as physics, chemistry, botany, as well as in most of the main lecture courses in medicine or law, or in mathematics at the technical universities, for example, the lecture fees often exceeded the salary, so that holders of professorships became wealthy.

This system, which was of questionable "morality," also had other unpleasant side-effects. The entire income of unsalaried adjunct professors and non-tenured associate professors consisted of lecture fees. Thus, only those from very affluent circles could afford to take such positions in disciplines with minimal lecture fees. Others who might take these positions were dependent on supplementary jobs, such as teaching at a secondary school. Another particularly demoralizing and significant effect was that the full professors usually took on the main lecture classes for themselves, in disciplines yielding medium to high lecture fees, leaving the secondary lectures or more advanced courses, which brought in very little, to their younger colleagues. Thus, the mathematics students in their first year at the University of Munich had to take classes mostly from aging full professors, whereas they would have been much more stimulated and motivated by the unsalaried adjunct professors. In the experimental sciences, including medicine, the negative effects were less drastic, since the unsalaried adjunct professors usually had fixed salaries as assistants, senior hospital physicians, or in other roles.

After the First World War, political considerations and the loss of many assets through inflation forced state offices to assure a minimal income also for non-tenured academics. This took the form of paid, temporary, part-time teaching assignments. Otherwise, the future of the universities might have been seriously threatened due to a shortage of qualified young academics. Experimental physics at the University of Munich was a typical example of the lecture-fee problem. As previously mentioned, on the one hand, the main lecture class, which was mandatory for most natural scientists and all medical students, was given by Röntgen, who was an inferior and boring teacher. On the other hand, however, the experimental physicist Leo Graetz, son of the Jewish historian Heinrich Graetz of Breslau, did not become a full professor (probably with a personal chair, *ad personam*, created specifically for him) at the University of Munich until 1908, when he was 52.[42]

Graetz was not a first-class researcher, but he was a brilliant teacher who could really captivate his students. He became known especially for his widely read textbooks and handbooks, translated into many languages, in particular those on electricity theory. Graetz taught a lecture class at the same time and on the same subject as Röntgen. It is not surprising that many students who registered for Röntgen's class actually went to hear Graetz's lectures. I remember that of Röntgen's 400 registered students, only about 100, and during carnival only about 50–60, were present at his lectures. Although many, especially medical students, simply did not go to the class, others regularly went to Graetz's lectures, without registering with or paying him. Sitting in on classes without paying for them, referred to as *schinden* (fee-dodging), was officially permitted only during the first two weeks of the semester. But if the professor did not object, as was the case with Graetz, who, on the contrary, was gratified by the students' recognition of his teaching talent, there was no penalty.

Röntgen did not concern himself very much with the relatively few students in mathematics, physics, and chemistry. However, he took the medical students very seriously, earning enormous sums from their lecture fees. He was also the examiner for the so-called *Physikum*, the first medical licensing examination after the fifth semester, although he found this additional paid job boring. As the medical students were well aware, Röntgen would begin by looking at the booklets listing the courses the students had taken, and woe to any of them who had taken physics from Graetz, for example, instead of from Röntgen! Therefore the students had to officially register for Röntgen's class. Incidentally, Röntgen was also an examiner for the doctoral and postdoctoral candidates in chemistry, and thus was feared also by their teachers, including even Adolf von Baeyer.

In my winter semester in Munich, I also took a lecture class on logic from a Catholic philosopher, since I had planned to take philosophy as my minor subject in the Ph.D. examination. I had no use whatsoever for the material taught in this lecture, which was given in an antiquated scholastic form. However, it made me appreciate Kant's historical misjudgment, according to which formal logic was

[42]Graetz had no connections to the Jewish community at that time. However, after the First World War he came closer to Judaism. For example, in the late 1920s, when the Nazis were already powerful in Munich, he presided over a meeting in which I told academics about Hebrew University.

supposedly a dying branch of philosophy, which could not develop any further. Quite the contrary, from the mid-nineteenth century and especially during the course of the twentieth century, the development and inherent potentialities of formal and symbolic logic were shown to surpass everything in the other branches of philosophy by far. Kant's misdiagnosis is all the more astounding since Leibniz already anticipated the later developments. "Symbolic" or "mathematical" logic has meanwhile become one of the most fruitful and widely applicable scientific areas.

When the summer semester at the University of Munich ended in late July 1910, I was pleased to join my mother and siblings, who were "running things" in Bischofswiesen. On hikes and relatively difficult mountain routes, I recuperated from the draining and not very stimulating first year of my studies, which, however, did not undermine my aspiration for a career in academia. Otherwise, I would have prepared for the examinations for a teaching certificate for secondary schools, and not somewhat frivolously studied, in 1911, the recently published book on the situation of non-tenured associate professors at Prussian universities. As a Jew, I could not even dream of becoming a full professor in those times.

In late October 1910, before the end of *Sukkot*, I went to Marburg University to start the winter semester. I was 19 and, as time would tell, my childhood ties to my family ended then. After that time, I returned to my home in Munich only for temporary visits.

Chapter 3
As a Student at Prussian Universities

For every young man of my background, the commitment to a religious (Orthodox) lifestyle and matters of sexual conduct, as well as the temptations presented by one's surroundings, gave rise to a number of problems.

Dr. Aron (Arnold) Barth, a close friend since 1912, used to classify Jewish boys with a religious upbringing into those who remained religious naively, without any crises, and those who were able to maintain their faith and traditional lifestyle only after going through serious confrontation with religious doubt. He preferred and felt closer to the former type, as demonstrated by his character and in his widely read publications.

I was different. As early as in secondary school, I had started a notebook entitled "Apologetics," in which I confronted doubt-inducing problems, jotting down actual or putative solutions that I found in literature or through my own contemplation. Even in school I occasionally gave voice to my doubts when, for example, Darwin was discussed, or Ernst Haeckel, who at that time was very popular among German youth. The teacher usually remained silent in such cases. Over time I limited my response in the following, perhaps all too convenient, manner: As a "natural scientist" I did not feel responsible for historical or text-critical problems, such as when I repeatedly read David Hoffmann's works on Leviticus and Deuteronomy. I learned a lot from these books written by such a profound, honest, and modest scholar. Borrowing several sentences from the introduction to Hoffmann's *Leviticus*,[1] I can say of myself that my "apologetic" stance was largely grounded in my strong desire to be able to remain committed to the tradition that had grown dear to my heart. On top of that came my father's statement to my brother and me that he would like his children to remain as religious as he had raised them, but I cannot say how much that influenced me.

---

[1]David Hoffmann, *Das Buch Leviticus*, trans. and explained, 2 vols. (Berlin: M. Poppelauer, 1905–1906). For an English translation of Hoffmann's "Introduction" to *Leviticus* see Jenny Marmorstein, "David Hoffmann—Defender of the Faith," *Tradition* 7–8 (1966): 91–101.**

A.A. Fraenkel, *Recollections of a Jewish Mathematician in Germany*,
DOI 10.1007/978-3-319-30847-0_3

As far as the common arguments against religion posed by the natural sciences are concerned, very early on I arrived at the certainty that there could be no conflict here, as they involved intrinsically different levels. Supposed contradictions, in my opinion, may arise from "religious" people wishing to deduce natural scientific facts from the Bible and other sources, for example about the age of the earth, but such facts cannot be found in religious literature. Or else natural scientists, in the sense of the familiar and often misused—because misunderstood—sentence by Pierre Simon Laplace in *Traité sur la Mécanique Céleste* (A Treatise of Celestial Mechanics), want to prove the nonexistence of God by reason of his nonappearance in science, for example in differential equations in physics. In fact, however, a natural scientist must be aware that there are forces and values that exist outside of observed nature and its scientific ascertainment, and indeed above it, and these might even represent the source of such nature. A scientist per se can neither affirm nor deny such forces.

Over the course of time I have defended my opinion in this regard verbally and in writing; in German, Hebrew, and English; to German-Jewish students and in the *Emek*[2] (valley) kibbutzim. I have encountered not only intense but also stupid dissent on both sides. For example, I indicated that in the story of creation, the expression "day" cannot have its common meaning because this account places the creation of the sun on the fourth "day." In response to my counter-question as to how *they* perceived "day" before the sun existed, my opponents, who accused me of heresy, answered, "Twenty-four hours of course." On the other hand, the famous Russian physicist Orest Danilovich Chwolson,[3] who was known especially for his textbooks, which were translated into many languages, wrote after Haeckel's *Riddle of the Universe* appeared that he read the book with "burning shame"[4]

---

[2]*Emek* may refer here to Emek Beit Shean, which has several religious kibbutzim. The kibbutzim in Emek Yizrael are secular.**

[3]Chwolson's father Daniel (Joseph) [or Daniil (Iosif) Abramovich Khvol'son] (1819–1911), born in Vilnius (then Russia), studied the Torah in Russia, Breslau, and Leipzig, and became professor of Oriental Languages at the University of St. Petersburg after being baptized in 1855. He was one of the leading Orientalists of his time. He spoke out to the Russian government many times and sometimes successfully in defense of the Jews and even after his baptism he maintained ties to Russian rabbis. He took up the sentence in the main prayer of *Yom Kippur* (Day of Atonement), whose ambiguity made it difficult to translate:

עד שלא נוצרתי איני כדאי, ועכשו שנוצרתי כאילו לא נוצרתי

Its original meaning is: "Before I (that is, the person in prayer) was born I was nothing; now that I have been born, it is as if I had never been born." As spoken by Chwolson it meant: Before I was baptized, I was a nothing, and now that I have been baptized, I conduct myself as if I had not been baptized.

[4]It was actually Friedrich Paulsen who wrote: "I have read this book with **burning shame** over the condition of the general and philosophical education of our people. It is painful that such a book was possible, that it could have been written, edited, sold, read, pondered, and believed by a people who possess a Kant, a Goethe, and a Schopenhauer." F. Paulsen, *Philosophie militans: Gegen Klerikalismus und Naturalismus* (Berlin: Reuther & Reichard, 1901), 187, as cited in Daniel Gasman, *The Scientific Origins of National Socialism* (London/New York: MacDonald/American Elsevier, 1971), 14. Similar in meaning, what Chwolson actually said was: "What feelings should

about how broad circles of the German people could embrace such a shallow and nonscientific resolution of the most profound problems facing the natural sciences and metaphysics. To my surprise I found an analogous stance also in very different Jewish circles. In a Palestinian kibbutz founded by Russian emigrants, my remark that modern physics raises serious objections to complete determinism prompted the response that: our worldview is deterministic and if modern developments in science contradict this, then we want nothing of such developments.

My religious worldview thus was grounded in tradition, but it never evaded the struggle between faith and doubt. My opposition to an "orthopraxic"—as opposed to an "Orthodox"—attitude, which was adopted by many religious academics in Germany after the First World War, was certainly too extreme. However, with respect to religious *practice* I have assumed a milder opinion over the course of decades. My earlier stringency, which even my father had occasionally disapproved of, surely often served to make things unnecessarily difficult for my wife, who was herself strictly Orthodox, during our first years of marriage, and of course also for myself, since my coworkers and superiors tended to resent my beard in general and my not shaving at all during the counting of the Omer and the "three weeks"[5] in particular. This development traces back in particular to a relaxation of formal legal attitudes (such as with respect to the consumption of dairy in countries in which the adulteration of milk is out of the question), but also to other moments such as the realization that many regulations, and even prayers, were interpreted in an ambiguous or cabbalistic manner, or when observing differences between Ashkenazi, Sephardic-Oriental, and Yemenite congregations as regards religious practices and liturgy. On top of that came my regret that in recent decades, and perhaps even longer, a growing and worrisome abstinence from any new *halachic* creations—except for some unnecessary complications—can be observed among rabbinical authorities, an abstinence that has catastrophic consequences in light of the needs of state functions in Israel and in view of the incongruity of certain liturgical texts once political independence was achieved. I will refrain from expanding this list, but would like openly to admit that I was and am certainly also influenced psychologically by the intolerance within Orthodox circles, starting with several previously mentioned experiences in Germany and moving on to unbridled outbursts of hatred. By that I do not mean those of the extremist Neturei Karta, which refuses to recognize the existence or authority of the "State of Israel," so such

---

these lines arouse in an expert of physics? Contempt or bitterness? What should one do? Should he laugh or weep?" and that Haeckel revealed "a scarcely believable lack of knowledge of the most elementary questions." See Armin Hermann, *Wie die Wissenschaft ihre Unschuld verlor* (Stuttgart: DVA, 1982), 78–79; cited with the English translation from Stefano Fait's Ph.D. dissertation in Social Anthropology at the University of St. Andrews (2004), and Chwolson, *Hegel, Haeckel, Kossuth und das zwölfte Gebot* (Braunschweig: Vieweg, 1906), cited in Helge Kragh, *Entropic Creation: Religious Contexts of Thermodynamics and Cosmology* (Hampshire/Burlington: Ashgate, 2008), 148.**

[5]The "three weeks" between the 17th of Tammuz and the 9th of Av, observed as a time of quasi mourning, especially the last nine days.**

actions could be expected, but for example those of the state-supporting Zeirei Agudat Yisrael against the Poalei Agudat Yisrael (PAI).

And now I'd like to make a few remarks about the problems in the area of sexuality that not only I faced, but also many young men with my convictions. A large segment of young, Orthodox Jewish-German men remained abstinent until they got married. This was not an exclusive prerogative of the Jews, but also had parallels in Catholic circles and there was even a Protestant student association that required its members to practice sexual abstinence until they married. This certainly required a great degree of willpower if matrimony was postponed until one's late 20s or early 30s, doubly so for students who sat next to attractive girls day in and day out in lecture halls, discussing scientific and ideological issues with them. And yet it was not as difficult as many people thought. For one thing, the self-discipline developed when living an Orthodox lifestyle and the consciousness that every sexual transgression is a serious sin are actually effective aids. And sublimation of the sexual drive through intensive intellectual efforts, through moderation in food and drink, and through a sensible selection of reading material and recreational activities, is easier than is often assumed.

In late October 1910, for the winter semester—the third semester of my studies, but my first one away from home—I went to Marburg in the Prussian province of Hesse-Nassau. I would have liked to have waited until after the High Holidays, but in 1910 (as in 1891, 1929, and 1948), the holidays fell at their latest possible date, and the math lectures in Marburg already began on October 25, the day marking the end of the festival[6] of *Sukkot*. Missing such a lecture, especially the first one, would have been inconceivable to me then.

Choosing Marburg for my first semester away from home was the suggestion of my uncle Alfred Loewy. Neither he nor I could have predicted at that time how significant this choice would be in many ways. His recommendation was determined—justifiably—by the older (though not yet 50) of the two full professors of mathematics, Kurt Hensel, who had announced for winter semester 1910–1911 that one of his classes would be a lecture class meeting four hours a week on number theory. Loewy had told me that, first, an "intermediate" university, and not Göttingen or Berlin, for instance, would be the most favorable fertile ground for me as a novice. And second, Hensel, although not among the foremost of German mathematicians, had proven to have a certain unusual originality with his odd, and at the time very much challenged, methods of algebraic number theory.

[6]This common name for *(Shemini) Atzeret* led in my final year of school to the following conversation. The Jewish New Year fell on Thursday and Friday, so that also in the subsequent weeks, except for the one immediately following the New Year, Thursday and Friday were holidays. In the last of these three weeks I said to the teacher on Wednesday that I would not be coming the next two days, whereupon he asked me: "Again? What holiday are you having now?" My answer was "the end of the festival." His retort: "Well it is high time that your festivals finally come to an end!"

Kurt Hensel

I remained closely connected with Hensel for 18 years, until 1928, and I remain greatly indebted to him. If he were aware of my depiction here of his family circumstances, he would not have admitted that it also presented a characteristic chapter from the final era of German Jewry. But after being forced to experience the Nuremberg race laws and the beginning of the world war before he died in June 1941, at almost 80 years of age, he could no longer deceive himself about it.

His background revealed the absurdity of the race laws in a grotesque form. The only so-called "Aryan" element in Hensel came from his paternal grandfather, the Berlin painter Wilhelm Hensel, who married Fanny Mendelssohn, the granddaughter of Moses Mendelssohn and older sister of composer Felix Mendelssohn Bartholdy. Fanny and Felix were both led to baptism by their father Abraham Mendelssohn. Kurt Hensel's own father Sebastian, who had turned to agriculture, is widely known as the author of the book *The Mendelssohn Family*,[7] which went through many reprints (18 by 1924) and which includes a description of the historic salons at Berlin's Leipziger Platz. Fanny's and Felix's younger sister Rebecka was married to P. G. Lejeune Dirichlet, one of the greatest mathematicians of his time and successor to Gauss at the University of Göttingen. Some interesting passages from the correspondence between the sisters can be read in Sebastian Hensel's book. Since Kurt Hensel's mother, Julie von Adelson, was also a baptized Russian Jew, he himself was considered 25 percent Aryan according to the Nazi legislation. For his children this percentage was lowered to only 12½ percent and some of his

[7] *Die Familie Mendelssohn 1729–1847, nach Briefen und Tagebüchern* (Berlin: B. Behr's, 1879); English: *The Mendelssohn Family,* trans. Carl Klingemann (New York: Harper, 1881).**

grandchildren were considered only 6¼ percent Aryan, since he married Gertrud Hahn, who came from a respected Jewish industrialist family from Berlin and was baptized prior to the wedding. Their oldest daughter Ruth married a baptized Jew, the University of Cologne law professor Franz Haymann, whose son, Walter Kurt Hayman, became an outstanding mathematician and professor at Imperial College in London.

Gertrud Hensel was a sophisticated, beautiful, and, despite her delicate health, energetic woman. Coming from a wealthy family, she was able to entertain guests with her husband in their palatial house atop Mount Schlossberg, high above the city of Marburg. I admired Gertrud Hensel very much, as did my wife later on. However, she was also the reason why Hermann Cohen did not socialize with the Hensels. In his later years Cohen refused on principle to enter a house if either the husband or wife were baptized Jews.[8] There was a peculiar convention regarding baptism that was not unusual in Berlin. It was common in some Jewish families, including that of the Hahns, that the academics among the sons—for example Martin Hahn, who later became a hygiene professor in Berlin—were baptized, as were daughters before they married a Christian. The brothers working in business or industry remained Jews, such as Dr. Georg Hahn of the Hahn steelworks, who later, when his company merged with the Hirsch Kupfer- und Messingwerke (Copper and Brass Foundry, HKM) company founded by Benjamin Hirsch, sat on the Hirsch supervisory board together with my father.

Hensel had written a significant doctoral dissertation in 1884 in Berlin while studying with the Jewish professor Leopold Kronecker, previously a successful businessman. Among German mathematicians at the time Kronecker had played an outstanding role, albeit one that definitely did not always have a favorable impact, and he also got baptized shortly before his death when he was 67 years old. The friendly relations between Kronecker and Hensel continued until Kronecker died in 1891. That same year Hensel became a non-tenured associate professor (*Extraordinarius*) in Marburg, though without payment, as Friedrich Althoff refused to pay him the salary he was due with the comment—as Hensel later told me—"You have enough money anyway." Starting in 1902 he was a full professor in Marburg. He turned down a professorship at Cornell University in Ithaca, New York, as well as one later on at the University of Leipzig, which came with the title of Privy Government Councilor conferred by the Prussian Ministry.

Toward the end of the century, Hensel switched his focus from the traditional areas of number theory and the theory of functions, which at that time received the most attention, to the subject that will always be linked to his name: p-adic numbers, a new creation based on imagination and intuition, which applied the methodology of the theory of functions to number theory. In this he initially experienced numerous disappointments. First, it was shown that he had conceived of his

---

[8]Cohen was on friendly terms with Kurt Hensel's brother Paul, a professor of philosophy at the University of Erlangen. He sometimes allowed Paul to tell him the latest Jewish jokes when he was visiting in Marburg.

idea—like many, except for geniuses the likes of Archimedes and Newton—without any solid foundation, which was then finally created in the second decade of the new century. Second, as a result of this he drew erroneous conclusions with respect to the problem of transcendental numbers, which were justifiably contested by Oskar Perron. Third, although he attempted in many publications to demonstrate that his method was significant for number theory, full proof of this was not determined until he was in his 60s by his highly gifted student Helmut Hasse.[9] Consequently, the impact of his new creation, while generally recognized today, was largely underestimated at the time. This was especially the case in Göttingen, which until the First World War and to a lesser extent until 1933 represented the Mecca of the mathematical world, attracting many talented young researchers from Europe and the United States. He was deeply embittered by this lack of acknowledgement.

Both from an objective scientific standpoint as well as subjectively, from that of my own development, I should add that Hensel's creation was to gain meaning in a different area, one which did not interest him much. In 1910, just when I went to Marburg, Ernst Steinitz's major work on the *Algebraic Theory of Fields* was published. The basic premise of this work had been influenced by the p-adic numbers and represented the cornerstone of one of the most important disciplines of modern mathematics: abstract algebra. I soon became more profoundly influenced by the thoughts of Steinitz than by the ideas that came directly from Hensel. Already in 1912 I turned to Steinitz with an extensive letter in which I sketched the basic ideas of the work that would later make up my doctoral dissertation and my postdoctoral professorial qualification (*Habilitation*).[10]

---

[9]For a largely scientific appraisal see H. Hasse's 1949 article: "Kurt Hensel zum Gedächtnis" (In Memory of Kurt Hensel), in volume 187 of the *Journal für die reine und angewandte Mathematik* ("*Crelle's Journal*"). The word *Jude* ("Jew") is not mentioned in the article!

[10]The system R of rational numbers $\frac{m}{n}$ has the property that the four elementary arithmetic operations (addition, multiplication, and their inverses: subtraction and division) can be applied without any limitations within the system, except for division by zero. R is therefore referred to as a *field*. Moreover, R is the *smallest* (infinite) field of numbers that is contained within a larger one, for example, in the field of all real (or complex) numbers.

However, there are also finite fields. Here as well, as in every field, it holds that a product is zero if and only if (at least) one of the factors is zero.

Already in the nineteenth century, more general fields had been considered, for example those whose elements are not numbers, but functions. An analysis of the most general "abstract" field and its properties was not conducted until 1910 by Ernst Steinitz, in his now classic 143-page treatise. As Steinitz himself stated, his analysis was prompted in part by "p-adic numbers," the innovative constructions (between numbers and functions), which were introduced by Kurt Hensel for very different objectives related to number theory and are too complex to explain here.

Such systems are called *rings*—especially when they contain zero divisors. "Abstract" rings were first examined in my dissertation and postdoctoral professorial qualification theses, and a short time later in a far more general sense by other researchers. Here too there is a connection to Hensel's number theory analyses, namely, to the g-adic numbers that appear in his book on number theory (1913), in which g denotes a composite number.

Ernst Steinitz

Now I must digress again from my chronological account to speak in this context about the Steinitz family, which clearly deserves recognition among the Jews of eastern Germany. Around the turn of the century and to some extent for a long time after that, the three highly talented Steinitz brothers lived in Breslau (now Wrocław in Poland). One of them, Kurt, whom I barely knew personally, was a famous lawyer. Another, Walter, had set up a practice as a doctor. He was actually more interested in zoology, and excelled therein to such a degree that at a mature age (he was 46 in 1928) he completed his professorial qualification in zoology at the University of Breslau. However, the Nazis' coming to power prevented him from being able to teach in the field. Although quite removed from Jewish tradition and scholarship, he was a Zionist and supported the establishment of a marine biology station in Haifa. When he later went to Palestine/Land of Israel, he was of course unable to get a position in his field since he lacked Hebrew proficiency and became a settler in the "German egg village"[11] of Ramot Hashavim. Walter later enjoyed the satisfaction of seeing one of his sons become a zoology professor at the Hebrew University.

The most eminent of the brothers, although the one whose life was most removed from Judaism, was the mathematician Ernst, born in 1871 in Laurahütte, in Upper Silesia. His great talent was apparent very early on, but as a Jew he did not receive a professorship until 1913–1914, at the newly founded Technical University in Breslau, after he had spent 13 years "starving" in a junior position at Charlottenburg College in Berlin.[12] A pure mathematician who was more interested

[11]Ramot Hashavim was founded by German immigrants, largely academics without university positions and lacking agricultural skills, who thus started chicken farms.**

[12]Charlottenburg College was the precursor to the Technical University of Berlin.**

in research than in teaching, he was not all suited for work in a polytechnic. It was his saving grace when he finally received a full professorship at the University of Kiel in 1920. There, he distinguished himself as a researcher in four apparently unrelated fields: topology, algebraic number theory, series analysis, and field theory. His seminal work of 1910 essentially founded abstract algebra. While today hundreds of researchers around the world work in the field of abstract algebra, it is characteristic of his time that the excellent mathematician Issai Schur, a Russian Jewish friend of Steinitz, told me in early 1913 that while the field theory work of Steinitz was interesting, he valued his work on algebraic numbers, published in 1912, far more. This is partly due to the fact that individual objects received greater emphasis in mathematics at that time, while today the focus is on systems of objects.

Later I will expound in greater detail on my relationship to Ernst Steinitz. In addition to Steinitz, among the founders of abstract algebra I should again mention Emmy Noether, whom I also knew well. A "nondenominational" Jew, she was dismissed from her position at the University of Göttingen in 1933 and spent the final year of her short but intellectually rich life teaching at Bryn Mawr College in Pennsylvania, where she is still remembered today.

While Hensel played a role in my decision to begin the third semester of my studies in Marburg, it was a good choice from a Jewish perspective as well. Despite the small population of both the town (roughly 20,000) and the Jewish community (about 50 families), which was formally Orthodox, though not very spiritual, Marburg had a kosher restaurant frequented largely by students. The rabbi, Dr. Leo Munk, had been one of the earliest graduates of the rabbinical seminary in Berlin. He and my father knew each other well as members of this institution's board. Before the First World War, and especially in the 1920s, there was a relatively high number of religious Jewish students in Marburg, about 15–20, mostly higher semester law and medical students. This was primarily because prior to the First World War, the University of Frankfurt (founded in 1914) did not exist, and because those who strove to work as lawyers in the Prussian province of Hesse-Nassau were required to complete their final semesters at the "regional university" in Marburg. The school of medicine did not, of course, enjoy the same status as that of the major urban universities, but the Protestant theology school, as well as the law school and the school of arts and humanities, which encompassed many subjects in the humanities and natural sciences, were very popular.

Although I got together daily with other religious students, I remained an outsider among the Jewish academics there. As I have done my entire life, I worked late into the evenings and went to bed early, with neither the time nor the inclination to develop my social life, except for occasional hikes or canoeing on the Lahn River. In order to get some physical exercise, I took a riding course which turned out quite disastrously—I was not terribly keen on riding in the first place and my instructor, who was usually drunk, terrorized me. Later in the Land of Israel/

Palestine, however, I got to enjoy riding. It enabled me to reach remote kibbutzim in the 1930s to give lectures.

I alienated myself from my fellow Jewish students mostly on account of my refusal to join any of the student fraternities. Almost all of the religious students belonged either to the orthodox Association of Jewish Academics, VJA, or—if they were Zionists—to the Association of Jewish Students, VJSt. However, I had an aversion to life in the fraternities and preferred to spend my time doing more important things. My Uncle Loewy also strongly advised me not to let myself be labeled from the outset by fraternity membership. It was a decision I never regretted.

Avoiding fraternities was just one factor in my not being very popular among my fellow students. It was intensified by my tendency, maintained throughout my life, to speak my mind clearly even if my opinions were unpopular. This came from my desire to remain intellectually independent and, in its somewhat crude and primitive Bavarian form, it brought me many influential enemies in my first few years at the Hebrew University. A characteristic example of this is my negative opinion of Samson Raphael Hirsch, who at the time was regarded as the virtual genius of the Jewish Orthodoxy in Germany and is still seen that way in certain Orthodox circles in the United States. From my early student days I declared this appraisal, both in private and semipublic settings, to be absurd. I rejected the symbolism that plays such an important role in Hirsch's exegesis, the dilettante linguistic explanations that lack any philological basis in the Semitic languages, the sentimentality of his ideological argumentation, especially in his *Nineteen Letters*, and his occasional ignorance. I saw a further shortcoming in his ambition to offer a—if not *the*—solution for all exegetic problems, an attempt that stands out regarding the psalms, instead of sometimes admitting the impossibility of such an explanation, as Rashi did. Because the positive elements of particularly the Pentateuch commentary are generally obvious to those familiar with the Talmud, I characterized Hirsch on occasion as the absolute "instructor for young girls," which evoked outrage among the students from the Frankfurt Secession circles.

In any case, these differences were essentially limited to objective areas, without getting very personal. I can illustrate this through a typical experience. Dr. Isaac Breuer, a lawyer from Frankfurt, was the third and most talented of Rabbi Dr. Salomon Breuer's six sons, that is, a grandson of Samson Raphael Hirsch, who is referred to as "Rabbi Hirsch" in Isaac Breuer's writings. He was a prominent "elder member" and *Fuchsmajor* (recruiter and advisor of younger members) in the Alliance of Jewish Academics, that is, the umbrella organization of all individual VJA branches. In 1911 or 1912 he gave a lecture for the Marburg VJA on a legal philosophical subject from an Orthodox Jewish perspective. I had been invited to attend and in the course of the discussion I made a number of critical, neo-Kantian comments, which were meant specifically for Isaac Breuer, a student of Rudolf Stammler, jurist of the Marburg school. While Breuer, who was known and feared among his friends for his caustic irony, sharply repudiated the other participants in the discussion, he responded objectively to me, the unknown guest, later speaking in detail with me and inviting me to his home in 1918, after the war was over.

Our friendly relations, borne of mutual respect, were reinforced in 1920 through our being related twice over, through marriages of our own families as well as those of our wives [see Fraenkel and Prins family trees]. That friendship continued despite occasional differences on Jewish and political issues until his death in 1946 in Jerusalem, where he emigrated in 1936. Isaac Breuer always made an effort to recognize and overcome his own prejudices. But it was not this virtue that formed the basis of our friendship. In his youth, up to around the end of the First World War, he was dogmatic, but he later developed an aptitude to learn from his experiences—no small matter for a man with his upbringing. It is characteristic of him that as late as 1932 he signed a petition to ban Orthodox schools in the *yishuv*, but a short time later became a patron of the Chorev school[13] in Jerusalem.

Let me now return to my teachers in Marburg. The second chair of mathematics was Professor Ernst Richard Neumann's, who in 1910 was only 35 years old. Neumann came from a well-known family of scholars, the grandson of the Königsberg physicist Franz Ernst Neumann and the nephew of the mathematician Carl Gottfried Neumann of Leipzig. As a scientist he could not hold a candle to their legacies, but he was regarded as a conscientious, if not particularly inspiring, teacher. It was typical of him that he lectured on the theory of relativity very early on, which as a mathematician he was by no means required to do. But he had recognized the importance of a subject that the Marburg physicists at the time found rather dubious and willingly put in the effort needed to familiarize himself and his students with it. The relationship between Hensel and Neumann, who had very different natures, both as individuals and as scientists, palpably worsened, though objectively speaking it was probably due to Hensel. After the war official communication between the two professors sometimes took a detour through me. Although I was not close to Neumann either personally or in our fields of research, he never made things difficult for me, even later when we were colleagues. Two typical episodes demonstrate his equitable attitude towards Jews, although he hardly had close contact with any.

[13]Chorev was established in 1934 by German Jewish Orthodox immigrants to Land of Israel/ Palestine. Founded with 16 boys and girls, it followed the religious philosophy of Samson Rafael Hirsch. Currently there are gender-segregated Chorev elementary and high schools.**

Hermann Cohen (Reprinted with permission from the National Library of Israel, Jerusalem)

Hermann Cohen (1842–1918), the well-known neo-Kantian and founder of the Marburg school, had in 1876 already become a philosophy professor in Marburg, succeeding his own teacher Friedrich Albert Lange, author of *The History of Materialism*. At that time the anti-Semitic wave had not yet reached the universities or the Prussian Ministry of Education and Cultural Affairs. However, before the turn of the century Cohen's manner had already fanned the smoldering anti-Semitic sentiments of his colleagues in the School of Arts and Humanities (*Philosophische Fakultät*) into flames. It went so far that many of his colleagues pointedly walked out when Cohen entered the professors' room at the university. When he retired in 1912 and moved to Berlin, where he taught at the Institute for the Study of Judaism (Lehranstalt für die Wissenschaft des Judentums), it was not enough for his enraged opponents to prevent his position from being filled by one of his students, such as Ernst Cassirer. They wanted to punish him by filling the now-famous philosophy chair not with a philosopher at all, but instead a teacher of experimental psychology. And in fact the chair was given to Erich Rudolf Jaensch, not yet 30 years old, who became known for his research on eidetic imagery and later for his Nazi attitudes. That marked the demise of the Marburg school, at least in Marburg. Neumann emphatically opposed this ideological decision, which brought him into serious conflict with the physics chair.

The other episode involved me directly, and I became aware of it from a letter that Neumann wrote in 1928, which I only happened to see in 1951. In the process of appointing chairs at German, especially Prussian, universities, it was common practice for members of the appointment commission named by the department to ask external colleagues to evaluate possible candidates. Then, after the list of suggested candidates was submitted to the ministry, the person responsible would contact some of the main authorities in the field. David Hilbert's student Otto

Toeplitz came from the "oldest" Jewish family of mathematicians in Germany. His grandfather Julius Toeplitz and his father Dr. Emil Toeplitz were secondary school mathematics teachers in Lissa (now Leszno) and Breslau, respectively, and he himself was a mathematics professor in Kiel from 1920 until the fall of 1927, when he assumed a position in Bonn. The appointment commission for his successor had considered me a possible candidate (and in fact I received the appointment); Neumann was therefore one of the people Toeplitz asked to appraise me, since he and I had been colleagues for 10 years. Neumann's lengthy response of January 17, 1928, is a prime example of objectivity. Although he did also mention the fact that as a physical mathematician he did not understand much about my main area of research, set theory, and found it too abstract and thus not very appealing, he did not forget to emphasize that I had "long become a recognized authority" in the field, followed by an extremely positive appraisal of my teaching experience, versatility, and character. On top of that he added that my lectures would have become even more popular had I not turned down the invitation to become a member of the examination commission for upper level secondary school teachers, "for reasons that I do not wish to discuss here." The reasons were clear to the writer of the letter and likely also to the addressee. Because those examinations in Marburg always took place on Friday evenings and necessarily involved writing, I had to turn down the invitation in order to keep the Sabbath, despite being well aware that this step would hurt me directly and indirectly. I had no idea what a favorable impression my staunchness would have on the German-national and conservative Protestant Neumann.

Toeplitz lost his chair in Bonn as a result of the Nuremberg race laws and went to Palestine/Land of Israel in 1938. Because he could not speak Hebrew, Salman Schocken offered him an academic-administrative position at the Hebrew University. As then rector of the university, I had the opportunity to work with him very often. He died in 1940. Not until 1951 did his widow, who was living in Haifa, find the letter from Neumann, which I had not even known existed.

Aside from the two full professors there were also two unsalaried adjunct professors (*Privatdozenten*) in Marburg at that time (1910–1914). One of them, an older and academically insignificant man, requires no special mention; but all the more deserves Ernst Hellinger (1883–1950), who came from a good Jewish mercantile family in Silesia. I met him for the last time in Chicago in 1946. After being dismissed from the University of Frankfurt, he had taken on a teaching position at Northwestern University. He did not publish much besides his extremely significant doctoral and postdoctoral professorial qualification theses in Göttingen on integral equations, but he was highly talented and a first-rate, conscientious teacher. I heard a number of his lectures. One of them (during the short summer semester 1911) on "set theory" proved to be decisive for my future, but not until a decade later. During his five-year stay in Marburg, Hellinger was virtually the director of *Crelle's Journal*, for which Hensel was the editor. Before I completed my professorial qualification Hensel asked me to take over this post, but I turned it down.

There is little to report about physics at the University of Marburg. I kept as much distance as possible from the then chair, an experimental physicist, because of his anti-Semitic sentiments. His successor (starting in 1920), however, the well-known Clemens Schaefer—who was very significant at a theoretical level and a devout Catholic—and I remained in close contact, even after I moved to Jerusalem and even after the war. The physics internship in Marburg, which was a requirement for my doctoral degree and which I had to endure for two semesters, although I hated it, cannot be compared to the standard of the internship I completed in the summer semester of 1912 at the Technical University of Munich, where Hermann Ebert was the institute director.

I have a lot more to say about Hermann Cohen, including some things generally not known. I contacted him right away since I had taken his large lecture class in 1910–1911 on "Kant's system," and then I also visited him at his home, when I brought greetings from an uncle of mine who knew him from the B'nai Brith lodge. The visit at his home was a failure: I nonetheless soon received a written invitation to dinner and, in response to my informing him that I keep a kosher diet, was told that only admissible "dairy" foods would be served. Rather awkwardly I also left that untouched on my plate. Although Cohen did not keep a kosher home, he wanted his Russian-Jewish students to keep kosher, and complained bitterly that they did not.

There is too little space to go into Cohen's original interpretation of Kant or into his own system, as set out in four substantial volumes, and no need to expound on his religious development. He was at first very remote from Judaism, but later identified religion with ethics and morality, and his late works ultimately affirm Jewish tradition and Jewish faith. He remained a perpetual opponent of Zionism. "The fellows just want to be happy,"[14] he used to say. The book he wrote after the outbreak of war on the similar essence of Germanness and Judaism seemed embarrassing and can only be explained as resulting from the war psychosis that affected broad circles of Jews at the time.[15] Other than sporadic contact with Rabbi Leo Munk, Cohen did not communicate with the members of the Marburg Jewish community, except for the congregation teacher Abraham Strauss (1865–1918). I too sometimes chatted with this unusually intelligent and highly educated man during long, inspiring evening hours. Aside from my Talmud courses with the rabbi, Strauss was the only person in the Jewish community with whom I had closer contact while I was a student. Cohen, however, was a friend of his former student, the Orthodox rabbi in Frankfurt Nehemiah A. Nobel. On occasion Cohen spent *Yom Kippur* fasting in Nobel's synagogue.

When in late 1910 I started attending Cohen's class in the main lecture hall of the University of Marburg, I was deeply impressed. Still today I can hear his citations

[14]This refers to a comment of Hermann Cohen to Franz Rosenzweig, criticizing Zionists of eudaemonism. Gershom Scholem said it was the "most profound remark that an opponent of Zionism ever made about the movement."**

[15]In this context it is interesting that at the reopening of the Frankfurt Free Jewish School on November 19, 1933, Martin Buber spoke of the German-Jewish (re)encounter of history. See Achim von Borries, *Selbstzeugnisse des deutschen Judentums 1870–1945*, Fischer Library, no. 439 (Frankfurt am Main, 1962).

from the beginning of Kant's *Critique of Pure Reason*. The severity of some of his personal comments did not bother me. He often criticized Spinoza and his concept of God, not only with philosophical arguments, but also with personal attacks; or he said that contemporary Berlin philosophy needed to be eliminated "*mit Stumpf und Riehl*"[16] (Stumpf and Riehl were two philosophy professors at the University of Berlin). After a few weeks I began doubting the soundness of his arguments. In his reference to the "facts" of mathematics and the mathematical sciences in particular, I missed any discussion of consistency. My distrust increased by virtue of his a priori arguments, which, in "transcendental aesthetics," were evidently supposed to rule out the existence of non-Euclidean geometry and similar well-grounded theories, and also because of the inconsistencies, as reported by Edward Study and others, in Cohen's own work, *Logik der reinen Erkenntnis* (The Logic of Pure Knowledge, 1902). Above all, however, I was deeply shocked at how infinity was treated in the Marburg school, from Cohen's *Das Prinzip der Infinitesimal-Methode* (The Principle of the Infinitesimal Method, 1883) to Natorp's *Die logischen Grundlagen der exakten Wissenschaften* (The Logical Foundations of the Exact Sciences, 1910), wherein the infinitesimal is brought into direct correspondence with Georg Cantor's transfinite numbers. A completely different, legitimate, and surprising defense of the actual infinitesimal was most recently—in 1960—achieved by a former student of mine, Abraham Robinson, who assumed a professorship at the University of California at Los Angeles[17] in 1962.[18]

---

[16]This is a pun on the German idiom: to eliminate something "mit Stumpf und Stiel" (root and branch).**

[17]Robinson moved to Yale University in 1967, where he remained until his death in 1974.**

[18]Here I am referring to Paul Natorp's book *Die logischen Grundlagen der exakten Wissenschaften* (1910) in order to promote understanding of the problem, not only because it is expressly based on Cohen's books on the subject (including Dimitry Gawronsky's Ph.D. dissertation), and follows them chronologically, but also because its presentation is much more understandable than Cohen's main work *Logik der reinen Erkenntnis* (1902; hereinafter: *Logik*), which was preceded in 1883 by *Das Prinzip der Infinitesimal-Methode*.

I will only briefly touch on the concept of space here, especially because a number of different interpretations of its treatment by Kant still exist today. The crucial point, aside from three-dimensionality, is the assessment of Euclidean geometry—in other words the characterization of the parallel axiom as a synthetic judgment a priori. It shall suffice to cite Natorp (p. 312): Euclidean space is "a 'necessary' presupposition in the specific sense that it is a condition for 'possible experience,' more specifically: for the unique (*eindeutige*), law-governed determinability of existence in experience." Even before the general theory of relativity, mathematicians and natural scientists had to reject this concept of space.

Far more characteristic for the Marburg school than the analysis of space tracing back to Kant is the emphasis on the meaning of the infinitesimal, that is, of "infinitely small" quantities, as first introduced by Cohen. Starting with Aristotle, and then in scholastic philosophy and theology, and more intensively since the seventeenth century, the infinite was discussed, but not until Georg Cantor's set theory did it acquire a firm foundation, in order to later play a key role in mathematics and a number of its applications. The Marburg neo-Kantians thus explained: If following Cantor an infinitely large "number" is denoted as w and if the ratio $1{:}w = x{:}1$ is formulated, then x must be an "infinitely small number." Such infinitesimal numbers could then be seen as, for example, numerator and denominator in the Leibniz notation $\frac{dy}{dx}$ for the differential quotient.

Aside from Cohen, the second professor of philosophy was Paul Natorp, 12 years younger than Cohen and a refined, musical man who was interested in teaching. Under the overwhelming influence of Cohen's personality, he had to pass himself off as a Kantian, specifically a logician, although this was contrary to his innermost self, which is why his publications lacked originality. After Cohen moved away (1912), Natorp blossomed and developed in accordance with his true nature.

In this context, Dr. Dimitry Gawronsky should be mentioned. A highly talented Russian Jew and revolutionary, who managed to escape to Germany after being condemned to death, he became Cohen's student around 1906. Cohen—supported in this by Natorp—declared Gawronsky the most talented of all his students. To be sure, this was not true; among the German-Jewish students he was certainly surpassed by Ernst Cassirer, and among the Russians by Sergei L. Rubinstein, with whom I maintained contact even after the First World War, until 1928, when he was the director of the university library in Odessa. Gawronsky had submitted his doctoral thesis in 1909 at the school of arts and humanities in Marburg, which Cohen judged as excellent. It was published in 1910 in Marburg as *Das Urteil der Realität und seine mathematischen Voraussetzungen* (The Judgment of Reality and its Mathematical Presuppositions). However, mathematicians in the same school at the university, especially Hensel, harshly criticized the work, quite justifiably in my opinion. It was ultimately accepted, but that of course served to intensify the antagonism. Cohen received little gratification from his German students in Marburg; shortly after he became emeritus, Nicolai Hartmann and Heinz Heimsoeth left his "school." I doubt that Hartmann even viewed himself as Cohen's student although Cohen was in fact his advisor for his postdoctoral professorial qualification. Regarding Natorp one cannot speak of "leaving," as he was captivated by Cohen. Once Cohen left, the magic gradually disappeared.

After I realized that I could not accept Cohen's philosophical doctrine, I had to decide if I really wanted to choose philosophy as one of my examination subjects. Since I was assured that Cohen wanted to hear his own views in the examination, rather than independent ones, I gave up the idea, without knowing at the time that Cohen would also be emeritus by the time I had my examination. Instead I chose (political) economics: first, because through my father's influence I had always been interested in the subject and even then it was related to mathematics via actuarial science—even more so today, as "econometrics"—and second, because the examination in that subject was especially easy in Marburg.

In reality, however, as science has indubitably recognized since the early nineteenth century, $\frac{dy}{dx}$ is not a quotient at all, but the limit of a quotient, and the "differentials" dx and dy definitely do not have the independent meaning that Cohen ascribed to them. However, he based his "principle of origin" on this as the foundation of his logic, asserting that with the infinitesimal method the "precise question" and "decisive answer" are thus given for the meaning of thought as production of existence, and dx is the "true unit," the absolute (*Logik*, 32, 109, 116, 123, etc.). In contrast, Cohen by all means rejects the limit method as the basis of the infinitesimal method—also for the integral. It is obvious that a mathematician must reject this view as false (or as meaningless).

In the fall of 1910 I started studying mainly mathematics. With the help of the Marburg rabbis, student lodgings were found for me with a Jewish family, where I could also prepare a simple dinner for myself. My express wish to have electric or gas lighting in the room was of course not satisfied until 1913 and I had to spend the winter evenings studying by the light of a petroleum lamp. I ate lunch at the Jewish restaurant, where the conversations and reports on lectures by fellow students tended to revolve around the unappetizing subject of jurisprudence.

I soon paid a visit to Hensel, having been introduced through a letter of my Uncle Loewy and the article on Easter that was published in *Crelle's Journal*, and was then invited to eat with them—on a Saturday afternoon. I therefore had to tell him of my Orthodox lifestyle. In conversation with "Mrs. Privy Councilor" a regular dinner date was arranged, and I was invited to a meal about once a month.

It was apparently my ritual-dominated lifestyle that prompted Hensel to read the memoirs of Pauline Wengeroff[19] and discuss them with me. This book by a well-to-do, tradition-conscious Russian Jewish woman does in fact offer some surprises for the uninitiated. Hensel did not even think of him or his family as having anything to do with Russian Jews, which is why he very innocently broached the subject. He painfully avoided identifying with German Jews in any way, despite the family pictures—for example of Moses Mendelssohn—in his salon. When he later told Jewish jokes to my wife he always changed the names, in hopes that no one would notice that it was referring to Jews.

My scientific interaction with Hensel took place on different levels, initially with respect to the number theory lecture class that was later combined with a seminar. I was most interested in Hensel's methodology of p-adic numbers in the case in which p is not a prime number, but a composite number g. On extended walks with me after the lectures, he would enthusiastically expound on the problems he was thinking about, although it was often difficult for me to follow his stream of consciousness. A third form of cooperation, which was important for both of us, came out of the plan he forged in 1910 to write a textbook on "elementary" number theory based on the g-adic numbers and therefore deviating substantially from all familiar textbooks. It was not easy for him to present his original thoughts in a form suitable for print, and that was precisely my strength. Because of the lecture class and our regular conversations on the subject, he asked me to put the book to paper (with the exception of the final, 12th chapter). His book *Zahlentheorie* (Number Theory) then came out in 1913.[20] This was the main reason why Hensel pressured me to extend my studies in Marburg to two, and then to three semesters, until spring of 1912. After my initial reluctance I soon recognized the important repercussions that this would have on my academic and professional career. Hensel suggested that

---

[19]*Memoiren einer Großmutter,* with a foreword by Gustav Karpeles, 2 vols. (Berlin: Poppelauer, 1908 and 1910) [English: *Memoirs of a Grandmother: Scenes from the Cultural History of the Jews of Russia in the Nineteenth Century*, translated with an introduction, notes, and commentary by Shulamit S. Magnus, 2 vols. (Stanford, CA: Stanford University Press, 2010 [vol. 1] and 2014 [vol. 2])**].

[20]Kurt Hensel, *Zahlentheorie* (Berlin and Leipzig: G. J. Göschen, 1913).

he be my advisor for my Ph.D. dissertation so I decided in 1911 to return to Marburg in spring 1913 for this purpose and then also to take the required economics lecture classes.

Meanwhile I had completed a major publication that appeared in 1912 in *Crelle's Journal*: "Axiomatische Begründung von Hensels *p*-adischen Zahlen" (Axiomatic Foundations of Hensel's *p*-adic Numbers).[21] This article brought me more success than it actually deserved. It was obvious that Hensel would value it highly, but in Göttingen people virtually said that it was through this article and not through Hensel himself that *p*-adic numbers were first irreproachably verified; that was not true. I offered a purely formal proof, which in my opinion was not all that difficult, while the actual substantial proof came a short time later (1913) from József Kürschák in Budapest and Alexander Ostrowski. My introduction to axiomatic considerations came from another source: my intensive participation in formulating and editing my Uncle Loewy's book, *Grundlagen der Arithmetik* (Foundations of Arithmetic), first published in 1915.[22] Working with him on this book had a lasting impact on me. The authors of these two books, Hensel and Loewy, also emphasized my participation in their prefaces.

A. Ostrowski

Alexander Ostrowski was a Russian Jew born in Kiev in 1893. He went to Marburg in 1911 since his professor in Kiev advised him to study with Hensel. Both Hensel and I immediately recognized the unusual talent and originality of the then 18-year-old who was not yet sufficiently competent in German. He and I remained friends until the end of my studies in Marburg, and much later our friendship was renewed in Switzerland. When the war started it meant hard times for him, as an

[21]In *Journal für die reine und angewandte Mathematik*, vol. 1912, no. 141 (Jan. 1912): 43–76.**

[22]Alfred Loewy, *Lehrbuch der Algebra*, vol. 1: *Grundlagen der Arithmetik* (Berlin: Veit & Co., 1915).

"enemy alien." In spite of his major research achievements, and although he completed his Ph.D. in 1920 in Göttingen and soon after acquired his professorial qualification in Hamburg and Göttingen, he could hardly have furthered his career in Germany. Consequently, he accepted a professorship at the University of Basel in 1927, where he remained for more than 30 years until he voluntarily retired and assumed emeritus status. I shared Marburg memories with him later in Jerusalem and in the beautiful house to which he retired in the Swiss canton of Ticino.

In addition to Ostrowski, in Hensel's lectures I also got to know the very talented German Jew Robert Remak, son and grandson of excellent physicians from Berlin and the nephew of Kurt Hensel. Three years my senior, from 1909 to 1932 he had numerous valuable works published, especially in the field of group theory and number theory. His arrogance unfortunately hampered his advancement, and he did not acquire his postdoctoral professorial qualification at the University of Berlin until he was 40 years old. After the Nazis took power he fled to Holland, from where he was deported to Auschwitz and killed.

During my five semesters in Marburg, I also took several classes from Neumann and Hellinger. Hellinger's lecture class on set theory made a great impression on me, whereby it was my good fortune that he mentioned the axioms upon which Ernst Zermelo of Göttingen had based set theory a short time earlier (1908). Zermelo's treatment of the subject would later become the point of departure for basic research on set theory, in particular also for my major research projects.

I took economics classes from Professor Walter Troeltsch (not to be confused with the famous theologian and philosopher Ernst Troeltsch). His lectures did not impress me all that much, but his courage to stand up to the mostly reactionary Prussian students did. The following wordplay will be understood by only few readers today: Professor Troeltsch opened one of his lectures with the words: "Now we will discuss the nature (*Wesen*), or rather the disaster (*Unwesen*), of the entailed estates."[23] This resulted in a deafening "shuffling" of the students' feet, through which they expressed their dissatisfaction with the professor, in contrast to "stomping," which was a sign of agreement. After the 1918–1919 revolution the entailed estates were dissolved.

Finally, I must briefly remember Walter Schücking, professor of international law. In addition to his normal, "private" lectures he also gave a *publicum*, a public lecture, in the university's main auditorium in the late morning every Saturday. This attracted a most varied audience—on the one hand political leftists and Jewish students who listened to the brilliant lectures with enjoyment, and on the other hand German-nationalists, who came with the intention of disrupting what they felt were unwelcome statements of the professor by shuffling their feet loud enough to drown

[23]Entailed estates are estates which cannot legally be sold or divided when inherited. Such conditions on estate can result in unfair division among heirs, or in heirs being rich in land but otherwise poor. When Professor Troeltsch said he wanted to discuss "the nature" of entailed estates, he added the prefix "Un-," his wordplay changing "nature" to "disaster," thus demonstrating his rejection of the concept of entailed estates and his courage in opposing reactionary laws and customs intended to benefit the rich.**

out the applause of the former group. Schücking used the *publicum* to oppose the prevailing reactionary political system in Germany, in particular Prussianism with its three-class franchise in the Prussian state general assembly (*Landtag*).[24] He was not ostracized among his colleagues as much as Cohen was, for example, because he himself came from an established Prussian family. After the First World War he remained in Marburg only a short time, when I got to know and value him personally, before he left to accept professorships in Kiel and Berlin. I became closer with him in the early 1930s when he was a member of the International Court of Justice in The Hague. From there he wrote me a sorrowful farewell letter in 1933, expressing his shame at the Nazis.

In summer 1911, when I was 20, I had to report for military service, which I dreaded. I turned up at the military doctor in Marburg who performed the routine examination and looked disapprovingly at my skinny body, displaying humor at my embarrassment at having to present myself undressed for the first time as an adult. Pointing to my glasses he asked what the matter was with my eyes. I replied, "Astigmatism." To which he blurted out, "That too?! *Landsturm*!" That meant that I would be in the reserve forces and serve active duty only in times of war. Despite the aggressive attitude of the German emperor at that time I naively thought a war was out of the question and was thus very pleased, thinking I had been freed from the military forever.

In early March 1912, after my third semester in Marburg, on my way back from Munich I made a stop in Heidelberg to spend several weeks attending a Talmud course given by Salman Baruch Rabinkow. I went largely at the suggestion of Fritz Gumpertz, whose letters in this connection I still have.

Rabinkow was born around 1882 in Sosnytsia, in the Chernigov Governorate. He was a student of Rav Eliezer Gordon (1840–1910) of Telz (Telšiai), but remained true to his Hasidic origins all his life. In the fall of 1907 he went to Heidelberg, initially to study philosophy and law. In view of his relations to several professors and as a result of the outbreak of war in 1914, he stayed in this university town for a considerable time. Not until the 1920s did he relocate to Berlin at the pressure of his students Dr. Fritz Gumpertz, Dr. Oskar Wolfsberg, and especially Dr. Aron Barth.

Two closely related aspects are significant regarding his importance as a *Talmid Chacham* (Talmud scholar) and his influence on his students in Germany. First, Rabinkow was a representative of the Lithuanian style of learning and thus was very different from those of the Hungarian style, which was clearly the predominant style among the great rabbis living in Germany. Second, he assumed an unusual and, according to serious evaluators, almost unique position among Lithuanian Talmud scholars.

Gumpertz also studied for a time with Dr. Heinrich Ehrentreu, one of the most eminent Hungarian Talmudists. As far as he and I could judge, a comparison of the

[24]In the "three-class franchise" system of Prussian representation, those eligible to vote (men over the age of 24) were divided into three classes based on amount of taxes paid, giving far greater representation to the rich.**

two styles revealed the following: The aim of the Hungarian method was the formal command of Talmudic material (including the commentaries and super-commentaries), the discovery of contradictions and the optimal strategy for their resolution, and an exploration of the historical and perhaps also the philological underpinnings of significant problems of interpretation. Rabinkow, in contrast, who was on a par with the Hungarian teachers with regard to his mastery of the material and the keenness of his memory, brought psychological depth to, and a more profound understanding of, the spirit of the Jewish law, which he categorized according to their uniform aspects without excessive dialectic refinement. Whereas the resources from Western scholarship upon which Ehrentreu drew were primarily of a linguistic nature, Rabinkow's approach was informed by his training in philosophy and even more so, in law. In fact, a number of Ph.D. theses at the University of Heidelberg's school of law were clearly influenced by Rabinkow. He was also familiar with modern perspectives in the natural and social sciences. His extraordinary talent for teaching, perhaps his most outstanding trait, can be largely attributed to his ability to present material systematically. When introducing a new subject he often began with a systematic and critical overview lasting hours.

If Rabinkow had any "weakness," it was his boundless drive for personal independence, to be at a remove from positions of authority. This correlated with his extreme material abstinence. All he needed was fresh fruit and salad or at most cooked vegetables. During the Third Reich he was forced to move to Holland. Many were surprised by news of his marriage. He died in 1942 from a serious illness.

The course I took from Rabinkow impressed me profoundly. It was here that I first became familiar with the Lithuanian *derekh limmud* (Talmudic method of learning) and then in such a distinctive form. I remained close friends with Rabinkow and in the 1920s was given the honored, yet ultimately futile, task of looking him up in Heidelberg and trying to convince him to become the director of the Berlin rabbinical seminary and to assume the Talmud studies chair. He valued his independence more than anything and turned down the offer.

I spent the summer semester of 1912 in Munich. This was when I was most closely involved with theoretical physics and attended a masterful lecture course given by Arnold Sommerfeld on thermodynamics and kinetic gas theory. The teacher of many great physicists, he gave a dramatic presentation in particular of Boltzmann's proof of the second law of thermodynamics based on probability theory. Sommerfeld, whom many (erroneously) considered Jewish on account of his physical appearance, was one of few in his field whose greatest achievements as a researcher came late in life. Aside from Einstein and Niels Bohr, no physicist impressed me as profoundly as Sommerfeld.

In Munich after the war I also met Sommerfeld privately, along with his Russian-Jewish student Paul Epstein, who was repeatedly my host in 1946 and in later years as professor at the renowned California Institute of Technology in Pasadena. When lecturing there I also had the honor of once having Robert Andrews Millikan and Carl Anderson in my audience.

Sommerfeld received a professorship at the University of Munich in 1906, where Arthur Korn, a Jew and almost the same age, had been a non-tenured associate professor (*Extraordinarius*) in theoretical physics since 1903. Korn was a distinguished and highly prolific researcher, also in the areas of electrical engineering and, like Sommerfeld, applied mathematics. Today's generation, which has forgotten or perhaps never knew, should recognize the name Korn as the inventor of the signal-conditioning circuits that led to the first television. His book on electrical photo transmission entered its second edition in 1907 and he published his findings in many German and foreign journals. The choice between Korn and Sommerfeld in 1906 might not have been an easy one, but in retrospect it must be acknowledged that it was the right decision for the University of Munich to transfer the chair to Sommerfeld. Still, Korn was offended and felt he had been passed over because he was a Jew; consequently, he resigned from his position in Munich in 1908. From 1914 he worked as an honorary professor at the Technical University in Charlottenburg (Berlin). One must keep in mind, however, that prior to the First World War there were very few full professorships in Germany in theoretical physics.

In early November 1912 I went into "Prussian exile" to spend the winter semester in Berlin. Although I did it very grudgingly, it in fact turned out to be the most wonderful semester of my studies, primarily from a Jewish and social standpoint, but also in terms of my studies.

I arrived at Berlin's Friedrichstrasse station around 11 o'clock in the evening; as I walked from there to my nearby hotel I was amazed at the crowds of people on the street at an hour at which Munich's streets were deserted—except during carnival. The next morning I started looking for an apartment and chose a modest room on Ziegelstrasse, midway between the university and the Orthodox heartland of the German Jews (Oranienburgerstrasse, Artilleriestrasse). I am embarrassed to confess that I barely got to know the centers of the Eastern European Jews. My "pad" turned out to be not entirely free of bedbugs, as other students living in this area also discovered. Sharing the same fate as us was Philipp Feist, the wealthiest in our student group, who had found a room in the "palatial" Bellevue district. Like many other students, I ate lunch in a kosher restaurant (Fifar)[25] near Oranienburger Tor. The eldest and also most distinguished of lunch guests that I met there was Joseph Carlebach, who later became the chief rabbi of Altona and Hamburg.

My social life revolved mostly around several families in the orthodox Adass Yisroel congregation, including David Hoffmann, Jakob Barth, and Joseph Wohlgemuth, who were teachers at the rabbinical seminary; but also other members such as Isidor Rosenblüth, representative of the Aron Hirsch & Sohn company and uncle of the longtime Israeli minister of justice Pinchas Rosen; and Sally Guggenheim, Rafael Gradenwitz, and Felix Struck, business friends of my father. I spent many pleasant evenings with my father's cousin Dr. Joseph Hirsch, a gynecologist. I also visited other Jewish scholars, such as the eminent Talmudist

[25]This refers to the lodge and restaurant at Friedrichstrasse 127, proprietor: Morris Fifar.**

Dr. Eduard Baneth of the Institute for the Study of Judaism (Lehranstalt für die Wissenschaft des Judentums). At these visits, and also later with Israel Lewy, the outstanding Talmudist from the Jewish Theological Seminary in Breslau, I became aware of the narrow-mindedness of the Orthodoxy. The two men were surprised and honored (although the honor was mine!) by this visit of a student of my background, the son of a member of the board of the Orthodox rabbinical seminary. And when I expressed my intention to attend external evening lectures at the Institute, a man as free of prejudice as Professor Barth said that I could not afford to, as it would give cause for unpleasant remarks. I was in the house of the distinguished classical philologist Hermann Dessau, whose brother Bernardo, physics professor at the university in Perugia, I already knew from St. Moritz. Bernardo Dessau had left Germany because he saw no possibility of receiving a professorship there; his artistically talented wife Emma was a sister of Rahel Straus (née Goitein), wife of Dr. Elias Straus.

David Hoffmann, longtime rector and leading Talmudist at the rabbinical seminary, was already 69 years old at the time. I remember from Friday evening meals with him how gently and modestly he would correct errors, and also how I kept pleading with him to complete his Deuteronomy commentary, indicating that the academic youth needed such works and not those of Samson Raphael Hirsch.

Friday evening dinners at the home of the great Semitic scholar and Arabist Jakob Barth was an even more enriching experience. Barth, who hailed from Baden, had been an instructor in exegesis and Hebrew at the rabbinical seminary since 1874. In 1880 he was also appointed a non-tenured associate professor[26] at the University of Berlin. His wife was a daughter of Azriel Hildesheimer, founder and first rector of the rabbinical seminary. The gentle, sentimental Hildesheimer must have been the ideal complement to the strict, and at times harsh, Barth. In addition to two daughters, the Barths had four sons, all of whom were already grown at the time; the youngest, Arnold (later Aron), later general manager of Bank Leumi in Tel Aviv, and I shared a close friendship starting in 1912, which became even closer when he moved to Palestine/Land of Israel in 1933, until his death in 1957. My personal friendship with Professor Barth, who always impressed me greatly, started back in spring 1909 in Merano, where I was recuperating from supposed mental exhaustion. Despite my only staying for two weeks, it was there that Alfred Loewy introduced me to infinitesimal calculus. Barth talked enthusiastically about his teacher Theodor Nöldeke and the doyen of Islamic studies Ignaz (or Ignác) Goldziher. Although Barth was orthodox and a member of the Berlin Secession congregation Adass Yisroel, he was hated by the Frankfurt Orthodoxy because he subscribed to the idea that the Book of Isaiah had multiple authors and that the psalms were written after the Exile.

Things were even livelier on Friday evenings at Wohlgemuth's, the junior lecturer in Talmud and later founder and editor of the *Jeschurun* journal. University

[26]At the beginning of the century he turned down a position as a full professor at a small northern German university. He was later conferred the title Privy Government Councilor.

and seminary students were usually there, including young girls, with whom I was forced into contact when I escorted them home. At that time it was considered dangerous for a young woman to walk alone, late at night, on the streets of Berlin. I was warned that the same fate could also befall a young man.

In Berlin, just as previously in Munich, I went to the city's outstanding theaters to see plays and operettas. Once my father's first cousin, Dr. Emil Hirsch of Halberstadt, invited me to the opera to see *Lohengrin*, but the performance was ruined for me when Emperor Wilhelm II appeared in the second act and was received with true Prussian Byzantinism.

In terms of academia, I did not accomplish as much during my Berlin semester as I had hoped. The reasons for this lay in the peculiar workings of the math department at what prior to the First World War was the largest German university. There were three professors of mathematics, Hermann A. Schwarz (1843–1921), F. Georg Frobenius (1849–1917), and Friedrich Schottky (1851–1935). All of them were in their 60s at the time and, with the exception of Frobenius, had long since passed their professional prime. Schwarz, son-in-law of the famous Berlin mathematician Ernst Eduard Kummer, had made his classic findings when he was much younger and then virtually stopped publishing. Schottky was a one-sided, old-school function theorist and a boring teacher. There was nothing that engaged me in either of their lecture classes. I also attended the "major" algebra lecture class given by Frobenius, one of the leading algebraists of his time, who remained productive into old age. As perfectly structured as his lecture was, it suffered from a crucial didactic shortcoming: It was so polished and apparently all-encompassing that students must have received the impression that this discipline was historically complete with no unanswered questions, which of course was the exact opposite of the truth. Furthermore, half of the lectures took place on Saturday mornings. For this reason, as well as from general inclination, instead of going to the later services at Adass, I went to the early morning services at the Bachurim Shul [i.e., synagogue]. There I sometimes had to read from the Torah ("leining"). I also met a lot of inspiring students from the rabbinical seminary, including the later esteemed rabbi of the Jewish Center of New York's Upper West Side, Dr. Leo Jung, whose father—London rabbi Dr. Meir Moritz Jung—had sent him to study at the Hildesheimer seminary. After the Second World War I also renewed this friendship from my youth in New York and Jerusalem.

The weekly two-hour mathematics seminar given by all three professors was structured rather oddly, so that each of them had a "turn" every three weeks. Frobenius, as he got older, was no longer very inclined to have the personal contact with students that such a seminar demands. In order to satisfy his obligations, he therefore posed a series of problems in the first class that were so difficult he could well expect that none of the students would even attempt a solution. Then, instead of a "seminar," he gave an in-depth lecture on more simple cases of Fermat's last theorem. Schottky gave a not very interesting monologue on his own research as it related to unpublished parallel work by Gauss. Schwarz took the most convenient route. He gave interested students complete freedom to report on whatever subjects they chose during the seminar. In unconscious anticipation of my future developments, I chose as my subject the elements of set theory, which was virtually unknown in Berlin. My two-hour lecture stimulated lively participation among the students. The professor then announced in his closing remarks that he too had heard about Georg Cantor's questionable theories, but felt he had to warn young students about them. In so doing he was following in the footsteps of Leopold Kronecker and not in those of his advisor and teacher Karl Weierstrass. He could not have had any idea at the time that only a few decades later set theory would be viewed by the majority of his colleagues as the foundation upon which all of mathematics was built.

As regards the rest of the mathematicians with whom I took classes at the university in Berlin, I would like to emphasize the class of Issai Schur on integral equations, as it was perhaps the best and most comprehensive mathematics lecture class I ever took. At the time the subject was hardly a decade old and analytical, whereas Schur's approach was primarily arithmetic-algebraic. But he knew how to tackle the subject in depth from all perspectives, in part using methods he himself had conceived, so that the few people in his audience did not only enjoy a work of art, but left having received a multifaceted mathematical experience.

Issai Schur

Schur was born in 1875 in Mogilev; his charming wife, Regina Malka Frumkin, was born in Kovno (Kaunas), which was still Russian at the time, and trained as a medical doctor in Russia. In their austere almost Puritan simplicity, and their incorruptible and sometimes critical attitude toward people and phenomena, they represented the best kind of Russian Jews living in Germany. I was often a guest in their home. Schur had gone to Berlin to study mathematics when he was around 20 years old. Residing at that time in the home of the retired Dr. David Hoffmann, he lived according to the religious attitudes with which he was raised. Frobenius soon noticed his unusually talented student and published many of Schur's articles on group theory, some of which they had written together, in the reports of the Berlin Academy (i.e., the Prussian Academy of Sciences). Nevertheless, even after he finished his professorial qualification in 1903 at the University of Berlin, he remained in the subordinate and unsalaried position of adjunct professor (*Privatdozent*), at first together with Edmund Landau. It was more pressing for him to receive a professorship than it was for German Jews, since according to Prussian administrative practices, Russian Jews were not naturalized. Just in the nick of time—though he and others could not foresee that of course—things changed when Edward Study's efforts led to his becoming a non-tenured associate professor (*Extraordinarius*) in Bonn (1913). Thus he became a public servant and automatically a Prussian citizen. Incidentally, when offered a position he returned to Berlin in 1916, and, deservedly, became a full professor in 1920 in Weimar Germany. When Ludwig Bieberbach and Erhard Schmidt replaced Schwarz and Schottky shortly after the end of the war, the mathematics department in Berlin was transferred into the hands of the younger generation. In late 1935, Schur too became a victim of the race laws; he spent his final years as an honorary member of the Hebrew University in Jerusalem, relatively happy in Tel Aviv, where his wife lived until 1965 and made good use of her Russian language skills.

Aside from mathematics, I also attended lectures in analytical mechanics by one of the greatest physicists, the then 54-year-old Max Planck (1858–1947), who is generally known as a Nobel laureate and father of quantum theory. Unfortunately, I felt almost the same way about him as I did about Röntgen. Although Planck's outstanding personality was clearly perceptible even to an outsider like me, I found his lectures boring. Admittedly, analytical mechanics is hardly as interesting as electromagnetism or even thermodynamics. Lectures on new subjects such as the theory of relativity or quantum theory were not yet common at the time.

My studies with Planck led to an encounter whose significance I realized only a long time later. The tutorials accompanying Planck's lectures were, as was customary, not led by the professor himself, nor were the exercises corrected by him, but by his assistant who at that time was a 34-year-old Jewish woman from Vienna, namely, Dr. Lise Meitner. Famous in later years as a radiologist and nuclear scientist, Meitner played a major role as a colleague of Otto Hahn in research into nuclear fission that later led to the development of the atom bomb in the United States. Aside from Meitner there were two other women involved in this research in the 1930s: Irene Joliot-Curie and Ida Noddack. Robert Jungk's interesting book *Brighter than a Thousand Suns*[27] tells about the dramatic developments in general and about the differences between Joliot and Meitner in particular. In any case, when I submitted my—certainly rather mediocre—exercises to Lise Meitner in 1913, I could not have known what a brilliant scholar was standing before me. After the Nazis came to power, her Austrian citizenship enabled her to continue at the Kaiser Wilhelm Institute, where she stayed until 1938. Once Austria was

[27] *Heller als tausend Sonnen: Das Schicksal der Atomforscher* (Stuttgart: Scherz, 1956) [English: *Brighter than a Thousand Suns: A Personal History of the Atomic Scientists*, trans. James Cleugh (New York: Harcourt Brace, 1958)**].

"annexed," however, even Otto Hahn's and Planck's interventions to Hitler himself were to no avail. She slipped over the border as a "tourist" and settled in Stockholm.

In spring 1913 I went back to Marburg to complete my doctorate within a year, without informing my parents of my plans. I had already independently prepared the concept of my dissertation and needed only to work out the details. The subject of the thesis was the theory of certain "rings," that is, abstract domains in which a product can equal zero although the factors of the product are all non-zero. Such rings in particular appear in Hensel's number theory. What I wanted to do was to postulate a general theory in which the elements of a ring do not necessarily have to be "numbers" as with Hensel's research. To this extent I was inspired by Steinitz's theory of abstract "fields"; certainly, a product in a field is zero if and only if (at least) one of the factors is also zero.

After a pleasant summer vacation with my family in Garmisch, in the Wetterstein Alps, I returned to Marburg in late October 1913 and, in addition to doing the final polish and refinement of my doctoral dissertation, I prepared to "hit the books" for the oral examination, which consisted of a two-hour exam in mathematics, physics, and political economics, which I completed in mid-January 1914 "with honors." According to my arrangements with Troeltsch, in economics I was tested in particular on the areas of finance and the railway system. Since I received the same rating for my dissertation from Hensel, who appreciated especially my independence, I had achieved my goal: to obtain my doctorate summa cum laude. This goal was determined not only by my personal ambitions, but in my case it seemed also to be an indispensable prerequisite for my academic career, that is, first of all for the postdoctoral professorial qualification (*Habilitation*). The obligatory publication of the doctoral dissertation did not cost anything, as the work appeared within a year in *Crelle's Journal*, with the title: "Über die Teiler der Null und die Zerlegung von Ringen" ("On Zero Divisors and Decomposition of Rings."). Franz Richarz, who held the physics chair, did not get along well with the two mathematics professors; however, he did not let his anger out on me, but on Hensel, by scoffing that in Hensel's department they now even "decompose zero."

On the Sunday morning following a Ph.D. examination, the doctoral candidate had to pay an official visit to his examiners dressed in a frock-coat and top hat. The liberal professor of political economics was very nice, but asked me anxiously if in view of my excellent examination scores I might be considering an academic career. In that case he felt in my own interest he had to warn me against such a step, since as a Jew there was no chance of my ever achieving a professorship.

As far as the professorship was concerned, we were in agreement and I did not need to contradict him. But regarding my academic career, the decisive step was taken only four days after my examination, with my having done nothing to further it. I had feared that there would be unsurpassable hurdles even to be able to complete my postdoctoral teaching qualification, so I was surprised and delighted when on the morning after my exam Hensel drew me aside in the university and told me that there were indeed prospects for me to do my postdoc research there. De facto there was only one unsalaried adjunct professor (*Privatdozent*) in mathematics at the University of Marburg who was needed for teaching. This adjunct professor was Ernst Hellinger. At that time—January 1914—the university in Frankfurt was about to open. It had been long planned, made possible mainly through generous donations by the Jewish businessmen and bankers of Frankfurt. The plans were for two full professors and one non-tenured associate professor (which was to be converted to a full professorship six years later) for mathematics, and Hellinger was supposed to receive the latter position. He had received the recommendation of his professor David Hilbert in Göttingen, the greatest contemporary mathematician; on top of that an agreement had been made with the Prussian Ministry of Education and Cultural Affairs that significant Jewish scholars, who due to their religion had been overlooked or had little prospects for the future, were supposed to be treated preferentially for Frankfurt. The most famous of all professors who received an appointment in 1914 in Frankfurt was the physician and chemotherapist Paul Ehrlich, Nobel Prize laureate of 1908. Up to then he had been the director of the Institute for Serum Research in Berlin-Dahlem and an honorary professor at the University of Göttingen, without, however, receiving a full professorial chair. His son-in-law Edmund Landau, on the other hand, had already become a full professor of mathematics in Göttingen in 1909. One needs to bear in mind, however, that the assessment of a mathematician's qualities as a researcher is less dependent on subjective factors than is the case in most other sciences. The Arabist Josef Horovitz, son of Frankfurt Rabbi Marcus Horovitz, is one of the most outstanding authorities on the Koran and he was not appointed to the chair for Semitic philology in Frankfurt until 1914.

In conversation Hensel told me about Hellinger's upcoming appointment in Frankfurt and of the need for someone new to receive qualification as an adjunct professor in mathematics in Marburg. He offered me the position of Hellinger's "successor"—of course after I had first continued my training elsewhere for several semesters and under the condition that I publish a suitably impressive postdoctoral thesis. As I noted earlier, an adjunct professor was not salaried and did not have a tenured position.

I was faced with the decision of a lifetime. As overjoyed as I was at the totally unexpected offer, I had a number of personal and professional considerations and also had to sound out my father on the matter. He came to Marburg for this purpose and had an extensive conversation with Hensel. Even before the semester had ended, in February 1914, I accepted the offer, while keeping open the question of the length of the interim period. While I had otherwise always maintained a student's thrifty lifestyle, in March 1914 I spent three weeks in Montreux recuperating from the eventful semester. No one could have imagined that this would be my last visit to Switzerland for many years. In light of the world war that started six months later, my doctorate and plans for my postdoctoral professorial qualification had taken on far greater significance. In the end this serendipity can be attributed to the energy my father had invested in facilitating my early entry into school, when I was just five years old.

For the summer semester of 1914, which unexpectedly became the final semester of my studies, I had trouble deciding between three very different places to study: Chicago, where two of the greatest American mathematicians in fields related to mine taught (Eliakim Hastings Moore, with whom I had already been in contact, and Leonard Eugene Dickson); Zurich, which brought together the outstanding Swiss Federal Institute of Technology (ETH) and the very good cantonal university, and which attracted me in any case on account of my great love of Switzerland; and finally Breslau, where Ernst Steinitz and other notable mathematicians taught. I cannot even hazard a guess as to how my life would have developed had I been in the United States, which long remained neutral, when the First World War broke out. Perhaps through inertia, and certainly not in response to any premonition, I decided on Breslau.

Although I did begin working on my postdoctoral thesis that summer, the semester was not very successful, either personally or scientifically. I soon started hating the city of Breslau, at first for external reasons: the environs, all the way to Zobtenberg (Mount Ślęża) or actually to the Riesengebirge (Karkonosze) mountains, are flat as a pancake and thus depressing. Also, despite enormous effort and numerous changes of address, I could not manage to find a bug-free room in the entire city center, where most of the Jews lived.

Jewish life in the community of more than 20,000 Jews was very developed, but also very splintered, which I found abhorrent. I visited a lot of prominent men in their homes. I have already mentioned Israel Lewy, who at the time was the leading Talmudist at the Jewish Theological Seminary. I also visited the rabbis of the main synagogues of the congregation, which had been unified since 1854, with orthodox and liberals enjoying equal status. Ferdinand Rosenthal was a Talmudist of distinction and Jakob Guttmann a significant philosopher of religion. I was particularly impressed by Guttmann and found his character to be even more orthodox than Rosenthal's. I had already met his son Julius Guttmann, one of the century's greatest Jewish philosophers of religion, in 1910 in Berchtesgaden. At the time he was an unsalaried adjunct professor of philosophy at the University of Breslau (as of 1919 he taught at Berlin's Institute for the Study of Judaism, Lehranstalt für die Wissenschaft des Judentums) and in 1934, as a professor for Jewish philosophy at the Hebrew University in Jerusalem, he became a friend and colleague of mine.

I cannot and have no desire to report on all of my Jewish contacts in Breslau, but I would like to mention one family that was typical of the extremes within the Orthodox community there: the Kober family. There were three brothers: Joseph, Wilhelm, and Aron.[28] Together they ran a successful metal company. Whereas Joseph prayed in the "White Stork," as the Orthodox synagogue of Rabbi Rosenthal was called, Wilhelm and Aron each went to a different "private synagogue," as there were numerous smaller Orthodox houses of worship in Breslau. The Jewish-political standpoint of the Kober brothers and their more or less grown children varied to an extreme degree: from "congregational orthodoxy" to separatism in the Frankfurt style, and from Zionist attitudes and practices—the daughters of Joseph and Wilhelm were Blau-Weiss (Blue-White) Zionist youth movement leaders—to extreme anti-Zionism.

[28] See Fraenkel family tree on page 204.**

I myself had a serious confrontation with Aron Kober at a public meeting. In the final years of my studies I had joined the newly founded Agudas Yisroel (or Agudath Israel), had even started a youth group in Munich, and had convinced Rebekka (Rivka) Feuchtwanger-Gluskinos ("Rhea Glus") to head the girls' division. She was from Breslau, and the wife of Dr. Feuchtwanger, later the vice president of Munich's Jewish religious community. When I was in Breslau, a meeting was organized to elect delegates for the first Knessia Gedolah of the Agudah (analogous to the Zionist congress), which then never took place because of the outbreak of the war. I demanded that in light of the diversity of the Agudath, members of the congregational orthodoxy and Zionists (Mizrachists) should also be eligible for election. Aron Kober opposed my stance with passionate attacks against Zionism. Unpleasant arguments ensued, which was all the more embarrassing for me since Aron's wife, Bertha, née Auerbach, was another of my father's first cousins and I was very close to her.

Noteworthy from a Jewish and Zionist perspective is another Breslau family. While I did not have far to walk from my "pad" to the university, the distance to the Technical College was about three miles. That is where the renowned mathematicians Steinitz, Gerhard Hessenberg, and Max Dehn taught. Actually, they had to adapt their curriculum to the needs of the engineering students and thus had little to offer me. But I really wanted to attend Steinitz's lectures and his pure mathematics lecture class was held on Saturdays from 10 to 12 o'clock. Even worse: The mathematics colloquium, which united the university and the technical college, took place every Saturday afternoon at the college. In order not to have to go back and forth several times following early morning services, that is, to have to walk a total of 12 miles, I looked for a Jewish family that lived near the Technical College, where I could spend Saturday afternoons and have time to relax. That is how I came to meet the Badt family.

Martha Badt—the hospitable and highly educated widow of the classical philologist Professor Benno Badt, who had taught at an academic secondary school—invited me to be a regular guest on Saturdays afternoons. Although only the younger daughter Lotte lived at home at the time, I also got to know the older daughter Bertha and the 27-year-old son Hermann. They were all Zionists very early on. Bertha, however, had married the decidedly anti-Zionist older son, Bruno, of the Marburg teacher Abraham Strauss, whom I mentioned earlier. Bruno taught at a secondary school in Berlin. Bertha was later widely known as the writer Bertha Badt-Strauss, especially after the two emigrated to the United States in the mid-1930s.

Hermann was a unique figure in the last German-Jewish generation. As far as I can remember, in summer 1914 he was a tourist in Palestine/Land of Israel, which tended to attract few tourists prior to the Balfour Declaration. From 1922 to 1926 he was a social democratic member of the Prussian state general assembly (*Landtag*). His extraordinary competence coupled with the absence of prejudice of the Prussian government at the time led to his being awarded the highest government office that a Jew—and an Orthodox one at that—had ever held in Germany. He became director of the Prussian Ministry of the Interior and Prussian representative in the

Reichsrat, the council of German states in the Weimar Republic. In this capacity he was a passenger on the zeppelin airship flight to the Middle East, which was the first time a dirigible ever cruised below sea level, that is, in the Jordan Valley near Jericho. It was Purim at the time and he read the *Megillah* of Esther in the balloon, surrounded by German dignitaries. After the Nazis took power, he had to flee Germany along with other political leaders including the Prussian prime minister Otto Braun. He and his family emigrated to Palestine/Land of Israel.

Lotte Badt was a medical doctor and still single at the time. She later married Dr. Joseph Prager, a neurologist from Kassel. After the war ended they settled in Breslau, where their house became a Jewish-Zionist center; from there they emigrated to Palestine as early as 1932. I always maintained contact with them and with Hermann.

The semester I spent in Breslau was, as I mentioned, not particularly productive in terms of my studies. The professors I was most interested in all taught at the Technical College, which was primarily for engineering students. The mathematics department at the university did not flourish until later, after the city became Polish. Nevertheless, even during my time there I received inspiring food for thought, especially from Erhard Schmidt, a Hilbert student from the Baltic states who rapidly made a name for himself and with whom I remained in contact. In the mathematics colloquium I presented my as yet unpublished doctoral dissertation, even though it fell on the Sabbath. It was a complicated matter, but I could not avoid it, and had reasonable success. Only Professor Dehn, whose baptism failed to accelerate his career as he had hoped and expected, commented that young mathematicians should not concern themselves with such abstract subjects, but should instead stick to concrete problems.

On June 28, a Sunday, I started out on a two-day trip to the Riesengebirge mountains with a son of Eduard Baneth. It was there that we heard the news of the assassination of the Austrian archduke and heir apparent Franz Ferdinand and his wife in Sarajevo. At first I did not grasp the fatefulness of the event for world politics, even less so as we did not have any idea at the time how ready and willing Wilhelm II and his general staff were to go to war. But in the course of July the storm clouds quickly gathered. I had made plans to meet a friend in Tiefenkastel on August 4, two days after the fast of Tisha b'Av (Ninth of Av), to hike together over the Julier Pass into the Engadin valley in southeastern Switzerland and then up into the Bernina range. My departure from Breslau went as planned, but the train ride was delayed due to military mobilization preparation and I arrived at my parents' just before the start of the Sabbath. When the Sabbath came to a close, not only had the day of fasting begun, but also a world war.

Chapter 4
As a Soldier in the First World War

The wealth of material I have about my experiences during the war could fill an entire book. Five bulging diaries, augmented by hundreds of military-post letters to my parents, offer a daily record from September 1914 to November 1918, with the exception of two weeks at the height of a serious illness in August 1917. An extensive account would, however, be uninteresting since there is no lack of literature on the daily experience and suffering of soldiers in the First World War.

My personal fate was at once easier and also more difficult than that of other soldiers. Easier, because I was never on the front line, and my contact with the enemy tended to be limited to aerial bombings.[1] Easier also because I was mostly engaged in professional service and could never complain of boredom, but rather of incredible tension, which is easier to endure. On the other hand, the difficulties connected with my traditional Jewish lifestyle, especially regarding food, but also concerning prayer, tefillin (phylacteries), not shaving, and many other things, led to a variety of additional problems. In fact I was always able to maintain my kosher diet, except for a short period in spring 1918, when I was specifically forbidden to do so in a letter from Dr. Ezra Munk, rabbi at Berlin's Adass Yisroel synagogue. It goes without saying that I was rarely able to observe the Sabbath.

[1] Incidentally, in 1909, following a lecture at the Munich polytechnic, I contradicted the prevailing opinion in Germany by predicting that future aircraft would be "heavier than air" and not "lighter than air." Later developments in aviation and the failure of the airships that were so intensely propagated by the German government before and during the war, especially the zeppelin, confirmed my prophecy.

A.A. Fraenkel, *Recollections of a Jewish Mathematician in Germany*,
DOI 10.1007/978-3-319-30847-0_4

Fraenkel as a soldier in World War I

The 50 months of my active service (1914–1918), following training as a medic in August 1914 with a final examination by Professor Julius Fessler, can be divided into two periods of roughly equal length.

Until late autumn 1916 I was on the western and southeastern front (Serbia) and then back in the west as a "medical orderly" (*Krankenpfleger*) in military hospitals. Although my position did not come with any prospects for military promotion, my duties in two different areas were more interesting, and carried more responsibility, than those of my comrades.

My superiors, rotating "army medical corps captains" and "army medical corps majors," felt I could be of far greater use by carrying out duties other than those normally done by a medical orderly. They had me transcribe autopsy reports from dictation; present case histories, including the anamneses (at times based on their instructions, at times based on my own observations); independently write the "monthly patient reports," which gave a complex monthly statistical record of the cases treated in all departments of the expansive military hospital; and assist in simple surgical procedures. The job that entailed most responsibility was the independent administration of ether anesthesia during operations, usually amputations and abdominal operations, but also isolated brain trepanations, which, when I was hesitant, I only performed after receiving clear military "orders." Anyway, I never experienced an error with the anesthesia, as did from time to time the "army medical corps staff sergeants," who were medical students who had not yet completed their studies. Thus I also gained all kinds of experience in internal medicine and pathological anatomy. My knowledge of Latin and Greek were very helpful in this area.

During the two years I worked in the military hospital, I was also authorized on the express orders of the Bavarian Ministry of Cultural Affairs to serve as a "Jewish chaplain." This did not relieve me of any of my usual duties, but it did offer some additional inspiration and gratification. The "pastoral care" consisted primarily of visiting Jewish patients in the various military hospitals, including those in epidemic control hospitals who were suffering from typhoid fever, typhus (spotted fever), or dysentery. I also had to conduct religious services in the field, especially on the High Holidays, for which Jewish soldiers from abroad were granted leave, by order of the army, within the scope of what was militarily feasible. I was even able to borrow some Torah scrolls for the service. A record number of participants—more than 150—turned up to celebrate the New Year (*Rosh Hashanah*) in 1915 in Cambrai, France. On the High Holidays in 1916 I was able to "*kasher*"[2] the field kitchens in Arlon and, with the help of religious cooks among the soldiers, offer the participants a meat meal; for the Passover *Seder* a "dairy" meal was provided for a small group of soldiers.

In September 1914, directly after the Jewish New Year, I entered Lorraine and spent the Day of Atonement (*Yom Kippur*) in a cattle car, that is, in a freight car that also in times of peace was used for so-and-so many people or so-and-so many horses. The day fell during the almost 48-hour transport of our and other formations from Lorraine to northern France. The hospital remained in Cambrai for almost a year. This was the easiest time for me. In late autumn 1915, after an interim period in what was then the Hungarian Banat region (Fehértemplom) and a railroad journey lasting more than six days to cover the roughly 30 miles from Versecz to the Danube crossing near the fortress of Smederevo, in freezing cold, without as much as a warm meal, the hospital was moved for the winter to the small Serbian city of Jagodina. A number of Sephardic Jews from Belgrade had fled there as the German army advanced. Whenever possible I would pray with them and they presented me with a Sephardic prayer book, which I still have, in which they inscribed their names. I also met Jews in the magnificently situated south Serbian (Macedonian) city of Üsküp (now Skopje).

[2]Make kosher.**

My Jewish activities sometimes brought me into contact with the military rabbis. There was usually one assigned to the High Command of each army, and, like the Protestant and Catholic army chaplains, they had officer's rank. There is nothing more to say than that the Catholic chaplains showed themselves to be most devoted to their cause. This was even more evident with the nurses: The Catholic nurses (nuns) were incredibly naïve as regards general education, but the welfare of their patients was paramount to them.

Fraenkel (sitting in the middle) among the medical staff during World War I

From the outset I had no illusions about the value of my "pastoral" activities, but on numerous occasions I was in fact pleasantly surprised. Many patients were grateful just to have a Jewish visitor. I felt the greatest satisfaction when afterwards some of the soldiers who had not had a religious upbringing wrote to me. Their letters were an indication of the impression that the events and our private conversations had on them. I know that this experience differs from those of many army rabbis. However, I had certain "advantages" over them. First of all, the sermon was not the absolute focus for me as it was for them. Starting in mid-1915, the subject of my sermons was increasingly the sufferings of the Jews in Poland, Russia, and Palestine/Land of Israel, and on the future of the Jews in general after the war. The emphasis of our religious services as a whole lay in the prayers and the *piyyutim*. In the exceptional and vulnerable condition of field duty, the familiar melodies from home had religious power that inspired trust and made it easier to carry on. That is why external participation and inner involvement were both greater. Besides—and perhaps this factor was crucial—the participants, hardly any of whom were officers, saw me as an equal and not as an officer who was privileged in every respect, as was the case with the army rabbis.

Many conscientious and excellently trained doctors, who were also distinguished in terms of their research, took care of the sick and injured. Yet I also saw cases in which the doctors' negligence claimed victims. In addition to the consulting internists and hygienists assigned to the individual armies, such as Professors Ludolph Brauer of Hamburg and Martin Hahn of Berlin, there were two exceptional pathologists, Professors Max Borst and Werner Hueck of the University of Munich (who later went to Leipzig), who conducted autopsies as well as surgical procedures. To this day I am proud of the honorable letters of recommendation that I received from them for my "assistance." No wonder they often did not get along with the acting director, who was primarily concerned with seeing to it that the beds were lined up in perfect order. I witnessed how one of these pathologists, in the middle of the daily surgical rounds, read that the next day he was relegated to run the typhoid ward (as punishment for insubordination). In response to his outraged objection that since his studies he had not seen a living typhoid fever patient, much less treat one, he was told that a military doctor had to serve as a specialist in all areas of medicine. Among my comrades in the military medical services, where many medical students worked, I can also remember individuals who later made a name for themselves in science, for example Bernard de Rudder, who worked in medical biometeorology, the borderline discipline between medicine and meteorology, a field that I always found particularly interesting.

Already in the early years of the war, scientific discoveries led to decisive developments. In the winter of 1914–1915, many soldiers died at the western front, suffering unspeakable torment from tetanus, including some who never received medical attention because their wounds appeared superficial. From that winter on, the treatment of the wounded with the tetanus antitoxin became mandatory in the German army, and the disease disappeared almost entirely. The same winter, tens of thousands of Austro-Hungarian prisoners of war died in Serbia of spotted fever (epidemic typhus). When the hospital I worked at was relocated to Serbia in late autumn 1915, people recalled that the disease could be transmitted only via body lice. Consequently, from that point on the troops were thoroughly deloused. All the same, in Jagodina we still had a ward for spotted fever, filled with seriously ill patients. The building had been completely deloused and the consulting internist demonstratively refrained from wearing the common white doctor's smock whenever he did rounds there, saying that the only place where he was safe from contamination was in the deloused spotted fever ward.

Many patients died in the dysentery hospital where I was a patient during the severe epidemic in eastern Galicia in August 1917. In fact they virtually gave up on me for two weeks. My weight dropped to 85–100 pounds. The doctor who treated (and saved) me called mine "a case to boast about." The rather conscientious ward physician knew nothing about treating dysentery and, in order to avoid doing anything wrong, had his patients eat nothing for two to three weeks, except for the dropwise ingestion of unsweetened tea or water. His request for red wine was rejected by the hospital director, who said that the wine was reserved for the officers' canteen.

It was thus no wonder that many died of exhaustion. I too suffered more from the indescribable weakness than from the actual pain.

From a Jewish political perspective, the war effected a major change in me. News of the fates of Polish and Russian Jews, and those in Galicia who were surprised by the Russian advance in 1915, had a profound impact on me. The expulsion of the Jews from the provinces of Courland and Kovno (Kaunas) particularly distressed me and really brought home the suffering of the Jews. It seemed to me that the war wanted to demonstrate the fate of a people without a country. Since I had long supported the idea of Zionism, I formally joined the Zionist World Organization and Munich's local Zionist group before being sent to the eastern front. I did not join the Mizrachi religious Zionist movement until a bit later, under the influence of Aron Barth. At that time in Munich quite an impression was made when a member of such a prominent and established family joined the local group in which Eastern European Jews predominated.

Meeting with Sephardi Jews in Serbia and especially, starting in spring 1917, with Galician Hasidim in Kraków, Lemberg (Lwów), Czortków, and other towns, greatly broadened my horizon, which had been limited to German Judaism. I was also deeply impressed by the Kraków synagogues (the High, Old, and Remuh Shuls), and the 400-year-old wooden synagogue in Chodorów. Despite its being prohibited by the military, for months I kept up frequent visits to a Hasidic mill owner, Shimon Weiler, and his family in the small village of Ottyniowice near Chodorów, southeast of Lemberg (Lwów), where the weather station of the German Army Group South had long been located. Weiler was not only a refined and deeply religious man, but also very knowledgeable in the Talmud. Through him and the information he passed on to me, the picture of Hasidism that I had acquired through Martin Buber's writings came to life, in expanded and modified fashion.

It seems as if Hasidic Judaism was preserved in a more original and pure sense in the villages, especially during the war—first under Russian and then German occupation—than in the cities. In Czortków, the famous Hasidic center where I prayed on the Jewish New Year in "Rebbe's Shul," the situation was not particularly satisfying. The Rebbe himself, who had run a real farm with great splendor, had long since fled to Vienna.

The following story is characteristic of the actions and the trust of Jews at that time: In early July 1917 the Russians under General Brusilov managed to break through the Austro-Hungarian lines near Kalusz, and the area around Chodorów was in danger of being surrounded by the Russian army. On July 13, a Friday, there was a mass exodus of the Jewish population fleeing to the northwest. After the Friday evening prayer in Weiler's home, the participants talked fearfully about the situation, especially if they should and could flee even though it was the Sabbath. Weiler himself, who had prayed on his own, came into the room during this discussion and shouted: "It is the Sabbath; this is not a time to have worries. Have confidence and carry on within the spirit of the Sabbath!" By the time the Sabbath ended the danger had passed, and the highly successful counteroffensive of the Army Group South began on July 18.

During the war I published several articles on Jewish themes in the Agudist and Zionist press in Germany, including "Jüdische Eindrücke im Feld" (Jewish Impressions from the Field), "Gedanken über künftige Entwicklung und Aufgaben der Agudas Jisroel" (Thoughts on Future Developments and Tasks of the Agudas Yisroel), and "Toratreues Judentum und Zionistische Organisation" (Torah-faithful Judaism and Zionist Organization). The last-mentioned article had the unexpected consequence that Hermann Struck, a painter I had not heard of at the time and then lieutenant and head of the Department of Jewish Affairs of the Ober Ost (Supreme Commander of German Forces in the East), wrote me an enthusiastic letter, which I received via the editors, since I had signed the article with my initials only. This letter led to a lifelong friendship, which became especially close in Palestine/Land of Israel. My Agudist article gave rise to a heated polemical conflict with the extremely anti-Zionist editorial office of *Agudah-Blätter* (Agudah Papers), which ended in my breaking with the Agudah. Outside the scope of this and other publications is the essay "Praktisches zur Universitätsgründung in Jerusalem" (Practical Ideas on the Founding of the University in Jerusalem), which appeared in 1918 in the third volume of Buber's monthly journal *Der Jude* (The Jew), and prompted a response by Harry Torczyner.

It might just be conceivable that I found leisure time while in the field and on leave for light reading. But I find it extremely surprising, even in retrospect, that I also found time for scientific activities that would one day be very significant for my future.

Already in summer 1914 I had considered the problems of my postdoctoral thesis, which would be a treatment of fundamental algebraic questions in the abstract "rings" that were largely first proven through my Ph.D. thesis. But one difficult question remained. Since work in pure mathematics actually requires only thinking, paper, and a pencil, and since it is known that soldiers spend most of their time waiting around, it was possible for me to resolve the problems in the first months of the war. The only remaining difficulty lay in formulating and writing down my findings, but it was helpful that for the first eight months of 1915 I remained stationed at one location (Cambrai). Although the military medical work was frequently more than abundant, there were still some free evening hours and also hours of waiting during the day. In short, toward the end of summer 1915 my postdoctoral thesis was ready to be submitted to the School of Arts and Humanities in Marburg. And by November, on Hensel's recommendation, it was already approved by the faculty.

Of course this was only the beginning of the procedure that included a public lecture, oral examination, printing of the postdoctoral thesis, and finally an inaugural lecture, before being granted the status of fledgling *Privatdozent* (unsalaried adjunct professor). I managed to complete the first two stages through a blessing in disguise. In early December I fell ill in Serbia with a particularly intractable flu and was released from the hospital in January 1916 and sent to Germany for R&R leave. Although the journey seemed endless and I was very weak, I was able to drive to Marburg while on leave, hold the lecture I had prepared while I was sick, went through my oral examinations, in which even my physics professor Franz Richarz, who had arrived in uniform, complete with war decorations, was merciful with the emaciated examination candidate. Before I returned to Serbia I sent my postdoctoral thesis to the publisher to be printed; my commanders were particularly accommodating and in 1916 granted me a short leave—this time from France—to go to Marburg, where I held my inaugural lecture on "Mathematics in the Actuarial Sciences" on July 12, 1916. In spite of the war I had nevertheless completed a major step in my academic career. I received my *pro venia legendi* ("permission to lecture," that is, my professorial qualification or *Habilitation*) in pure mathematics and actuarial mathematics.

While this first piece of success can be attributed to careful planning, diligent execution and a modicum of good luck, my second scientific achievement during my military service was more like a lottery jackpot: conceived by chance, it became one of my life's greatest successes.

While on leave in the fall of 1916, lying awake one night in bed, I thought about what I would do with any future leisure time I would have in the army. I thought of Hellinger's lecture on set theory in 1911, whose subject had so completely captivated me that from time to time I engaged in conversations with army comrades, introducing them to the world of infinite numbers. Why shouldn't I write a popular, comprehensible little book about it and then get it published after the war? No sooner said than done! While still in Munich I started writing the first pages. Even though my meteorology training and getting settled in the Galicia field weather station took up all my time at first, starting in the summer of 1917 I frequently had a few free hours between observations and calculations in which I could continue writing my book. It was a project that not only came easily for me, but it also inspired me and distracted me from unpleasant military experiences. The book was completed in the summer of 1918. I sent the manuscript first to B. G. Teubner in Leipzig, which had long been the leading mathematics publisher in Germany. It was returned quickly, with a rejection and a rather implausible explanation from an expert consultant I had never heard of. Disappointed but not yet discouraged, I sent it—also from the field—to the Julius Springer publishing house in Berlin, an up-and-coming new academic publishing company, which only a few years later would take over Teubner's role in the area of mathematics. This time I was lucky: The expert consultant was Leon Lichtenstein, a significant Jewish mathematician from Poland, who at the time was still an unsalaried adjunct professor (*Privatdozent*) at the Technical University in Berlin-Charlottenburg and later became a full professor at the University of Leipzig. Even later, we remained friends.

The modest little 156-page book appeared in 1919 and proved very popular.[3] A third edition, by then a scientific reference book of 424 pages, was published in 1928 as volume 9 in the famous "yellow-cover" series under the direction of Richard Courant (official series name: *Grundlehren der Mathematischen Wissenschaften*, Comprehensive Studies in Mathematics). It was reprinted during the Second World War in the United States. The book was even more widely read after I changed its format, rewriting it in English and dividing it into two parts: the

[3]A. Fraenkel, *Einleitung in die Mengenlehre* (Berlin: Julius Springer, 1919).**

first, elementary part had its third edition in 1965.[4] A much shorter Hebrew edition was published in 1953. This was dedicated to the memory of Berl Katznelson, with whom I collaborated closely on many questions of adult education. And so the work that was conceived quite by chance and first written under the most primitive of circumstances ultimately led to sales of more than 10,000 copies, not only in Europe, the United States, and Israel, but also in Japan and in Russia, where it had been translated without my knowledge. Although inferior to some of my short research papers in terms of scientific value, the book was my *magnum opus*, contributing primarily to my career and making my name familiar in the world of science.

In December 1916 I was detailed to the "Kogenluft LEA." This mysterious name is an acronym for the Commanding General of the Air Service, airship units (**Ko**mmandierender **Gen**eral der **Luft**streitkräfte, Luftschiffer-Ersatz-Abteilung). Initially I had to spend some time as a recruit in the airship troops barracks in Reinickendorf near Berlin, in order, on the basis of my civilian qualifications, to work in the weather service. My penchant for meteorology and the recommendation of Professor Siegmund Günther, who as a captain was the director of a weather service, led me in this direction. My time in the barracks, during the coldest months of the so-called Turnip Winter, so named because an early frost coupled with the supply needs at the front line meant there was little else for civilians to eat, bring terrible memories to mind, which can only be described by words such as corruption, theft, and bribery. In any case, at that time, too, I successfully avoided having to learn to shoot. I was overjoyed when I was detailed to the "Main Weather Station of the Homeland" in Jüterbog, where a more scientific and less military atmosphere prevailed. Our training was led by Professor Emil Less, who was already over 60, a *Privatdozent* in meteorology at the University of Berlin. There were several gentlemen trained there before and with me who were more significant scientists than Professor Less, including Alfred Rühl, the later professor of geography at the University of Berlin; Professor Hans Ludendorff, brother of General Erich Ludendorff; and Robert Wenger, later professor of geophysics at the University of Leipzig. The latter, who sadly died very early, made the most lasting impression on me. He was a student of the Norwegian Vilhelm Bjerknes, rightly regarded as the father of the modern science of meteorology. Instinct and experience, the two factors that played a crucial role in forecasting weather at that time (and perhaps still do today), surely could not get their due in the context of military training and practice. On the other hand, in my many conversations with Wenger, we discussed manifold difficulties regarding data analysis, which at that time were considered insurmountable, but in today's age of electronic computers no longer exist.

After completing my training I was stationed in early April at the Field Weather Station of the German Army Group South in eastern Galicia. In July 1917 I was promoted to officer candidate (*Gefreiter*) and in late 1917 to Prussian corporal

---

[4]A. Fraenkel, *Abstract Set Theory* (Amsterdam: North Holland, 1953); see also A. Fraenkel and Y. Bar-Hillel, *Foundations of Set Theory* (Amsterdam: North Holland, 1958).**

(*Unteroffizier*). A short time later I received the rank of a "scientific assistant" and had to provide the necessary services either under the chief of the weather station or in his stead. This was actually a lieutenant's position, but since according to regulations it could be filled if necessary by a corporal, it seemed to rule out promotion even to the rank of sergeant (*Feldwebel*). Nevertheless, my rank as corporal gave me a sense of satisfaction that I simply cannot describe.

Duties at an army weather station like ours were multifaceted and interesting. The main tasks were to make regular meteorological observations day and night, including the upper-level winds. The weather maps had to be drawn three times a day from aggregate dispatches gathered from dozens of weather stations throughout Europe and transmitted in numerical codes by telephone. It was also necessary to prepare and transmit prognoses for the weather in general and the upper-level winds in particular for the army. These obligations were easy for me; I had more trouble with the fact that I had always slept very lightly and suffered from insomnia prior to extraordinary events. That became a serious problem for me. If I did not have to record observations between midnight and three in the morning myself, then I was awakened by my comrades getting up during the hours of my deepest and most needed sleep.

In addition to the daily responsibilities came special tasks. For example, a neighboring *Minenwerfer* (trench mortar, literally: mine launcher) training school requested a lecture series on the meteorological conditions for launching Minenwerfer gas shells. Because the director of our weather station did not feel competent enough to do it, he ordered me, the *Gefreiter*, to give the course. My students were almost all officers and when at the end of the course the commander of the school, a captain, wanted to thank me, he didn't know how he should address me and began with "*Euer Hochwohlgeboren*."[5] The humor of this form of address can only be appreciated by someone who served in the German Army.

From Jüterbog I had already signed up for the weather service of the Turkish army, which was allied with Germany. My main motivation was my hope that this would be a way to be sent to the Land of Israel/Palestine as a soldier. Another consideration was my prospects, as a scientist, of quickly being promoted to Turkish officer.

In early August 1917, when I had already endured attacks of fever, the Army Group South advanced. The Army High Command was supposed to relocate to Czortków, to the palatial home of the Rebbe, and the weather station also started moving. In Brzezany, where the environs had been completely shot to ruins, we encountered a group of Jewish civilians. They were refugees, and the young female vendors with whom I could talk in Hebrew complained that the German soldiers wreaked even worse havoc than the Poles and Ruthenians before them, whereas since the revolution it was surprisingly the Russians who were addressing the situation of the Jews in the occupied parts of Galicia and Bukovina. Then we

[5]"Your Highly Well Born" is a form of address for members of the lower German nobility or landed gentry.**

continued on farther through Buczacz (home of Agnon[6]) to the village of Wygnanka, high above Czortków. There were so many flies in this region that one could easily kill thousands in one room without noticeably reducing their numbers. We had barely arrived when I came down with dysentery. The disease was evidently transmitted by the flies and had remained latent during the arduous advance. My symptoms of course worsened visibly as a result of the poor nutrition and lack of treatment. In the days that followed, the disease spread epidemically among the soldiers in the area, making a special dysentery hospital necessary. The severity of my case meant I was one of the first patients to be admitted. During the weeks I spent hovering between life and death I received notice from General Headquarters that I was being transferred to the Turkish weather station, with instructions to travel to Constantinople. The doctor reported back to the General Headquarters that illness and weakness rendered me unable to embark on a journey—either presently or for an extended period of time. That was the end of my dream of serving in the Land of Israel/Palestine. At the same time a request came from the Prussian Ministry of Education and Cultural Affairs for me to be given leave for four months during the winter semester 1917–1918, since I was needed at the University of Marburg. "Kogenluft" refused, however, on account of the shortage of meteorologically trained personnel.

When the eastern theater of the war was demobilized in early 1918, we were regrouped and sent westward. As of March 1918 I served at a weather station set up in a railway car not far from St. Quentin. It did not work for the local army there, but specifically for a bomber squadron whose commander treated me harshly. This was tied to completely new duties. In addition to the upper-level winds, the main task was to determine prognoses of ground fog, as most airplane crashes—especially at night, when the bombers flew—happened not through enemy action, but when landing in fog. I was also given an essentially mathematical task: to calculate the course direction necessary for the bombers, especially toward Paris, taking into account the expected wind direction and wind strength at high altitudes. I wrote a treatise on this and on the maximum possible error, entitled "Kursberechnung für Flugzeuge" (Course Calculation for Airplanes), which was published for the German army by the General Headquarters.

As is generally known, the major German Spring Offensive of March 21 approaching Amiens ultimately got bogged down. It had been the last victory of German weaponry, which was largely possible in the first place on account of the meteorological conditions and led to an advance of about 45 miles, something virtually unfathomable at the time. The retreat began in June with the second Battle of the Marne. The enemy tanks and American ordnance proved to be the deathblow for the Germans. I experienced very little of it, as I fell ill again in July and afterwards was stationed temporarily at the Bavarian main weather station in Munich. All there is to report is that our commander, Lieutenant August Kühl,

[6]Shmuel Yosef Agnon, 1966 Literature Nobel Prize laureate.**

observer at the Munich observatory, treated his subordinates, including several scientists, with military severity and harshness.

I was supposed to move out as an assistant at a field weather station in early November. Four days before that, on November 4, I heard an impressive speech together with many thousands in Munich. It was on a par with that of a statesman, by the great economist and politician, Professor Max Weber. He was later pushed to the sidelines by the shortsightedness and egoism of the Democratic Party functionaries, instead of being given the opportunity to play a decisive role commensurate with his abilities in the newly founded Democratic Party and thus in the new German republic. Then things started happening very fast. On November 7, two days earlier than in Berlin, revolution broke out in Munich after unforeseen multitudes attended a popular assembly on the Theresienwiese grounds. The Bavarian parliament was occupied by the council of soldiers, workers, and farmers, which established itself as the new authority after the king and the government had been ousted with very little bloodshed.

The days were extremely eventful and varied from another perspective. My Zionist sentiments would be expressed in action. After many preliminary conversations I managed to found a Mizrachi group in Munich on November 10. On November 11 I had to return to the barracks where the weather station was located. Extended discussions took place. I spoke out strongly for moderation, but against all militarism. And I wrote a letter—signed by the soldiers' council—to the Ministry of War, in which we demanded that Kühl be forbidden from entering the barracks. In the days that followed I received my leave and left immediately for Marburg to start teaching.

Chapter 5
As a Professor in Marburg and Kiel (1919–1929)

My first visit to my parents' home as a civilian took place under exciting external circumstances in the spring of 1919, after I had finished teaching my first semester in Marburg. After Kurt Eisner, Minister President of the Free State of Bavaria, was murdered on February 21, 1919, the USPD (Independent Social Democratic Party of Germany) and the communists, who were active under a number of different names and who included some Jews, seized power in Munich and especially in southern Bavaria. On the night of April 6, 1919, the Revolutionary Central Council of Bavaria declared the Bavarian Soviet Republic. Political murders were the order of the day in Bavaria's opposing political camps. When I arrived at Munich central station that March, the train could not enter the station and the passengers had to take cover from the machine guns in the streets. The Passover festival took place under increasing terror and worries about my father, who, as a religious Jew and representative of the mercantile community, was being attacked from both the left and the right. The climax came in April: The dictatorship of the Bavarian Soviet government, made up exclusively of communists as of April 14, led to the murder of ten hostages on April 30 by the so-called Red Army in the Luitpold Gymnasium, my former secondary school. On May 1 government troops from northern Germany marched into Munich. I was an eyewitness to the ensuing street fighting with cannons. On May 2 Gustav Landauer was stoned to death by soldiers. A short time later, militant anti-Semitism became visible in Bavaria. By early May, however, trains started running again and I was able to return to Marburg.

A few weeks later I met my future wife, Wilhelmina (Wilma) A. Prins, in the home of her sister Meta Posen[1] in Frankfurt am Main. Her Hebrew name, Malka, means queen. She was her parents' seventh child and it was common for Dutch Jews to name the seventh child after the king or queen of the Netherlands—at that

[1]See Prins family tree on p. 205.**

A.A. Fraenkel, *Recollections of a Jewish Mathematician in Germany*,
DOI 10.1007/978-3-319-30847-0_5

time it was Queen Wilhelmina. My wife was and is the ideal partner for me, precisely because our natures and interests—except for religion and Zionism—were almost opposites, though in many respects they complemented each other. Wilma lived with her parents in Amsterdam and had majored in German studies at the local university for several years. Both her grandfathers were Jewish scholars: Liepman P. Prins of Arnheim-Amsterdam, who later moved to Frankfurt, and Rabbi Marcus Lehmann of Mainz, who was known mostly for his Jewish literary work. I did not fail to warn Wilma prior to our getting engaged about the dismal material and political situation in Germany and the unfathomably different situations faced by Jews in Holland and Jews in Germany.

We got engaged in September 1919 in Amsterdam and married there shortly before Passover, in late March 1920. At our wedding table I gave a speech in Hebrew, in which I vowed in both our names to dedicate our lives to the Jewish people, if possible on the soil of our historical homeland. My wife was a Zionist, the only one among eight siblings, long before I was.[2]

Precisely on those days in mid-March, shortly before the wedding, my warnings to my wife about the uncertain conditions in Germany proved not to be unfounded, as the Kapp Putsch almost made it impossible for me, my parents, and my siblings to travel from Marburg and Munich to Holland.

As in the rest of Germany, in Marburg there was an extraordinary housing shortage such that the owners of large apartments were legally required to take in "compulsory tenants." When in late fall 1919, I introduced my fiancée to the Hensels, they suggested that we move into the upstairs of their home, in compliance with the tenancy ordinance. This had some disadvantages with respect to Jewish and household considerations, but overall the suggestion was ideal and we gratefully accepted the offer. I moved from my "student pad" to the Hensels' home under indescribably difficult circumstances and almost entirely on my own, since it was virtually impossible to arrange for help in view of the political unrest and the nationalist putsch. Over the course of April the situation stabilized somewhat.

[2]Two weeks after our wedding, a brother of my wife, George Prins of London, lost his wife Viva shortly after she gave birth. Viva was the only daughter of Berthold Israel, the well-known Berlin department store owner, whose younger son Wilfrid was a philanthropist and social worker in Germany, England, and Palestine/Land of Israel. Wilfrid died on June 1, 1943, while flying from Lisbon to London, when his plane was shot down by the Germans. It was through the influence of George Prins—as a British government liaison and member of the Diamond Syndicate—that the diamond industry was founded in Palestine/Land of Israel in 1939–1940.

The Hensel villa in Marburg, where the Fraenkel family resided with the Hensels

Our presence turned out to be less than ideal for the Hensels. After our eldest daughter Rahel was born in November 1921, our first son Benjamin Ze'ev (Wolfgang) followed in May 1923, our second daughter Tirza in July 1925, and our youngest, Aviezri Sigmund (named after my father) in May 1928—all in the Hensel house! Instead of a quiet married couple, a family with four noisy children was squeezed into the tight quarters. By early 1928 the situation became unbearable for the Hensels, and I started a desperate search for another apartment. Just at the right time, in March, I received a professorship in Kiel.

For the fall and spring holidays we generally went to either Amsterdam or Munich, where we and our children were spoiled in every way. I would like specifically to emphasize the role played in this respect by my in-laws, father-in-law Maurits Prins, who lived until late 1932, and my mother-in-law Emma, who in 1942, when she was already over 80, perished in the Bergen-Belsen concentration camp, where she had been deported along with other relatives who had remained in Holland. From 1921 to 1926 we always spent the month of August as guests of my parents in Reichenhall, which was our starting point for hiking tours in the Berchtesgadener Land region as well as for shorter trips to Tyrol and South Tyrol (Alto Adige, in northern Italy). My wife increasingly grew to enjoy mountaineering, although she had not been accustomed to it in youth. We went to Switzerland on our "belated" honeymoon in August 1920, to Zermatt and the Engadin; from 1927 on, Switzerland became our regular destination insofar as we could afford a summer vacation. Thus our children became familiar with the Alps very early on; and in 1931 they even got to know the Grindelwald glaciers.

As a *Privatdozent* (unsalaried adjunct professor) and a non-tenured associate professor, my meager income came only from lecture fees that the students paid. Despite the generosity of our parents, we still had to maintain a very frugal household—with the exception of 1928–1932. The same was the case later on in Palestine and Israel until 1954, when we started doing well financially. My wife, who had grown up in affluent conditions, felt the economic pressures of our new way of life more than I did, but she was able to adapt to the situation both in Germany and in Jerusalem.

For the first time in late March 1921 we visited Wilma's oldest sister Selina in London for several weeks. I had known her rather well from Munich and St. Moritz, from the time before the First World War. She was married to David Solomon Sassoon. Although Selina was the actual lady of the house, she and David were living in the Sassoon household with David's eminent mother Mrs. Flora Farha Sassoon. Both mother and son were extraordinary people, each in his or her own way. They came from the famous Baghdad-Indian mercantile family and had moved from Bombay to London in 1902. David had left the family business very early on, in order to collect and publish as a private scholar. Flora Sassoon was adept not only in secular matters, but was also very knowledgeable in the Talmud and rabbinical literature. It once happened that while sitting at the table I got into a dispute with her over an interpretation of a Talmud passage and she sent her young grandson (who grew up to be Rabbi Solomon Sassoon) to get the appropriate volume of the Gemara, in which she showed me a *tosafa* that supported her opinion. She was also given the honor—something unique for a woman—of presenting the yearly opening address at Jews' College in London, a speech well-documented with quotations from the Talmud and *Midrash*. She also spoke in her home with Indian dinner guests, including maharajas and aristocratic Muslims, as well as with the house personnel, who were mostly from India, in Hindustani, Arabic, and other Asian languages. The grandchildren did not speak to their grandmother, father, and private tutors in English, the colloquial language in the house, but in Baghdadi Jewish Arabic.

David Sassoon, a well-to-do and highly erudite scholar, had accumulated a unique library of Hebrew and Samaritan manuscripts on his travels in the Middle East (Iraq, Syria, North Africa) and India. He worked for more than a decade on a scientific catalog of these manuscripts, which was published by Oxford University Press in 1932 under the title *Ohel David*, in two large quarto volumes with magnificent plates. The work contains a wealth of new literary discoveries, whose scope far exceeds that of the actual catalog. I was lucky enough to be able to view many of these unique manuscripts when visiting my brother-in-law, such as the oldest extant Pentateuch manuscript, the "Farchi Bible," and the edition of the *Mishnah Seder Mo'ed* with its Arabic commentary which was written and used by Maimonides himself.

From right: Wilma Prins Fraenkel, Selina Prins Sassoon, Abraham HaLevi Fraenkel, David Sassoon

It might be interesting for me to report something about my brother-in-law's illustrious family and family history, especially since they were renowned far beyond Jewish circles. Sasson ben Saleh ben David ben Jacob (1750–1830), leader of the Jewish community in Baghdad, was the "founder" of the Sassoon family and father of the "first" David Sassoon 1792–1864).[3] Father and son were forced to leave Baghdad due to the persecution by Da'ud Pasha. The son continued on to India, where Jews from Baghdad had already settled in the early eighteenth century. He started a business in Bombay, which was the point of departure for what became the widely branched trade and financial empire of the Sassoons, who were considered the "Rothschilds of the Orient" into the twentieth century. They, and the Baghdadi families related to them by blood or marriage, including the Somekh,

[3]Sassoon or Sasson is a Hebrew word meaning "joy"; it was and still is used as a first name in the Middle East.

Gabbai, and Ezra families, played a decisive role in the development of trade in the entire region of the Middle and Far East as far as China. The first David Sassoon financed and built the trading docks in Bombay that, to this day, bear his name.

In addition to Bene Israel and Cochin Jews, Baghdadi Jews soon made up the third Jewish community in India. Economically they became the most significant by virtue of their financial strength, which was primarily concentrated in the companies of Sassoon. The first David Sassoon erected synagogues, schools, hospitals, and different kinds of philanthropic institutions at many locations in India and elsewhere. The later British First Commissioner of Works (1937–1939) Sir Philip Sassoon and Siegfried Sassoon, the English poet born in London in 1886, were both great grandchildren of that first David Sassoon. However, by no means are all Sassoons in Israel and the Middle East related by blood or marriage to the family discussed here.

Suleiman (Solomon) Sassoon, son of the first David, was born in 1842; after the death of his father he became one of the leaders of the firm, but he died early at the age of 52. His widow Flora, who was equally distinguished in business, Jewish scholarship, philanthropy, and by virtue of her social standing, survived him by more than 40 years. They had a son, David, who later became my brother-in-law, and two daughters, Rachel and Mozelle, the latter of whom was ill and died in the 1920s. Rachel married the philanthropist Sir David Ezra in Calcutta, whose achievements included the founding of that city's famous zoo. Lady Rachel Ezra was highly educated, both in Jewish and secular matters. I often met her—sometimes together with Sir David—in England, Holland, and France prior to the Second World War. We kept up a correspondence till 1950.

My brother-in-law David, who was born in 1880 and lost his father only three months after his bar mitzvah, received a very thorough Jewish and secular education. For the former, his mother had tutors come to Bombay expressly from Baghdad. As mentioned earlier, he left the family business very early in order to dedicate himself exclusively to Jewish, scholarly and philanthropic matters. Characteristic of his genius as a collector of manuscripts, he recognized that a manuscript that had been rejected by the British Museum was in fact part of a Mishnah commentary written by Maimonides himself and, of course, acquired it for his own collection.

Mother and son left the famous Sassoon house atop Malabar Hill in Bombay in 1902 to move to London, where they lived on Bruton Street in the West End until the Second World War. They were the only branch of the Sassoon family that remained true to Jewish tradition. In 1912 David married Selina Prins, my wife's sister, who is now over 80 years old.[4] They had one daughter and one son.

On their frequent trips to the European continent, the Levant, and North Africa, staying in the most varied luxury hotels, the family was always accompanied by their Indian-Jewish cooks, a married couple who prepared Indian-style kosher meals, just as they had in London. Of David's many characteristic traits, at least one is worth mentioning. While he was on one of his research trips in Kobe (Japan)

[4]Selina Prins Sassoon (1882–1967) died at the age of 84.**

as a young man, a Jewish woman with no known acquaintances died suddenly. David saw to it that she received a ritual burial, including soil from the Land of Israel. When he was later asked, "Where did you get the soil?" he responded, "Since my bar mitzvah I never travel anywhere without taking my burial shroud and some soil from the Land of Israel with me." Many such biographical details demonstrate his touching humility and piousness.

Among his many publications—aside from his aforementioned monumental work *Ohel David* and the posthumously published *A History of the Jews in Baghdad* (1949)—I should mention his travel diaries, written originally in English and then completed and translated by Meir Benayahu and published in Hebrew in Jerusalem in 1955 as מסע בבל (Travels to Babel).[5]

Starting in 1933, Mrs. Flora Sassoon, David, and Selina saved many German Jews by providing a livelihood for them in England. At the time of the "Blitz" (1940), the family relocated their residence to the "Garden City" of Letchworth in Hertfordshire, where many Orthodox Jews from London settled for the duration of the war. David died in 1942, and Selina continued to live in Letchworth with her children and grandchildren.

Our first of many visits to the Sassoons in London was in 1921. That visit, which was also my first trip to England, showed me the English civil service from a typically congenial side. When after a stormy ferry ride at night from Holland we arrived seasick in England, we were refused entry despite the fact that our visas had been correctly entered in our German passports, because people were suspicious of "enemy aliens." With much effort we convinced the port authority to contact the authorities in London, but, nevertheless, we had to return to the ship with a notice that we would have to travel back to Holland the following night. Meanwhile I had been questioned by a police officer who wanted to verify my professional data. He asked me . . . about squaring the circle! Although he was weak in mathematics and at the time I was even weaker in English, the interrogation was carried out in a very gentlemanlike way, and he was satisfied with the result. A short time later the Home Office permitted our entry. In 1928 David and Selina came to visit us in Marburg, as the *Gevatter*, or godparents, of our youngest son.

This is not the place to report in great detail on my scientific work between 1919 and 1929.[6] However, I would like to make a general remark, which has only recently become widely known outside of the scientific field. Creative talent in the exact sciences—mathematics, theoretical physics, astronomy, and symbolic logic—is essentially reserved for the youth, from about 15 to 35 or at most 40 years of age. The few exceptions prove the rule. For surprised readers let me first offer three typical examples, a list which could easily be expanded many times over.

---

[5]The Hebrew edition (312 pp.) was published by Azriel Publishing. It is also available online at http://www.otzar.org/wotzar/book.aspx?149421. An excerpt was translated into English by Rabbi Aharon Bassous, in *The Scribe,* no. 75 (autumn 2002), 20–24: http://www.dangoor.com/TheScribe75.pdf (accessed July 16, 2016).**

[6]Anyone interested should refer to the Festschrift: Yehoshua Bar-Hillel (ed.), *Essays on the Foundations of Mathematics*, dedicated to A. A. Fraenkel on his Seventieth Anniversary (Jerusalem: Magnus, 1961, and Amsterdam: North Holland, 1962).

Évariste Galois at age 15

First, some of the greatest mathematicians of all time died before they turned 30; the Frenchman Évariste Galois (1811–1832) died in a duel at 20. Certain of his imminent end, the night before the duel he wrote a last scientific will and testament, the deciphering and implementation of which occupied the mathematical world for more than half a century. Among his brilliant contemporaries were the Norwegian Niels Henrik Abel (1802–1829) and the German Jew Ferdinand G. M. Eisenstein (1823–1852).

A second example, which is often cited as counterevidence by those unfamiliar with the subject, is the greatest mathematician of the nineteenth century, Carl Friedrich Gauss (1777–1855), who is known to the world for an achievement he considered insignificant—the discovery of the electrical telegraph.

Carl Friedrich Gauss

During his long lifetime Gauss published unceasingly in the fields of mathematics, astronomy, and physics. His collected works comprise 12 large quarto volumes.

Some of his greatest discoveries were never published and were finally found posthumously in his estate. A scientific diary discovered long after he died surprised posterity, showing that all or almost all of Gauss's exceptional achievements were made when he was between 19 and 37 years of age, whereby he also mentioned typical evidence of his talents at the ages of 3, 10, and 15. Most of his later publications of great significance are elaborations or developments of thoughts he first had in his youth.

The third example is Einstein. As I have previously said, the highlights of his creative work took place in the first decade of the twentieth century. His peak achievements, belonging to very diverse areas of physics, were published in 1904–1905, when he was 25. He had just completed his doctorate in Zurich, and would still have years to wait for his first academic position. Einstein did not achieve anything outstanding after he turned 34. Despite his unremitting efforts to solve a highly complex—and possibly unsolvable—problem, he was ultimately unable to make a decisive contribution.

It is beyond the scope of this book to analyze the phenomenon of "early aging" in mathematics and its methodologically related fields. Physiological reasons certainly play a crucial role. A creative mathematician or composer produces "something from nothing," that is, through contemplation and imagination alone. A historian, in contrast, or a literary researcher, and in particular also a biologist, relies on given material, whereby maturity and experience play a greater role than the creativity of a young brain. I do find it tragic that mathematics and theoretical physics have become such widely branched fields that the creative spirit at the optimum age—around 20—has not yet had the time to gather all the knowledge necessary to develop important new ideas.

I have mentioned these remarks, which readers might judge to be a pointless digression, in order to emphasize a danger that appeared to threaten me after the world war, though at the time I hardly noticed it. The advances in abstract algebra that are contained in my doctoral dissertation and my postdoctoral professorial qualification thesis were interrupted by my more than four years of military service during my "prime," from age 23 to 28; later I no longer had the strength in me to further develop my innovative thoughts. This was subsequently done by others, especially Wolfgang Krull. Other relatively extensive works that I published after the war, about Gauss for instance, contained no true originality.

Just in the nick of time, when I was 30–31 (1921), I conducted two different investigations in a new field (foundations of set theory) and presented papers on my findings at the annual meetings of the German Mathematical Society and the Göttingen Mathematical Society in 1921 and 1922. These papers were published in renowned journals in 1922.

I don't know if it was plain luck or an unconscious inkling, but I selected my two main research areas—abstract algebra and the foundations of set theory—at times when almost no one paid any attention to them, so it was relatively easy to achieve significant results. Both became key areas of interest for mathematicians within only a few years and that interest has continued without a break to the present day. I already discussed the first area. As far as the axiomatic foundations of set theory are concerned, Ernst Zermelo[7] had published a seminal work in 1908, when he was in Göttingen, and I heard about it in passing in Ernst Hellinger's lecture class. Although little attention was paid to it for 13 years, it is still rightly regarded as a classic today. In my *Einleitung in die Mengenlehre* (Introduction to Set Theory) of 1919, I had already devoted several pages to that publication by Zermelo. It occurred to me that Zermelo's axioms were not sufficient in forming certain sets—a discovery that also surprised Zermelo himself. I therefore introduced a new axiom schema of "replacement" and sent a short article, which also included other improvements to Zermelo's system, to the *Mathematische Annalen* (Mathematical Annals), where it was immediately accepted by David Hilbert.

Ernst Zermelo

[7]Although they are extraneous here, I would like to mention a few practically unknown traits of this brilliant and peculiar mathematician, whose name retains an almost magical resonance even today. A poor teacher, he did not get far in Göttingen, even though he had published a three-page article in 1904 proving the "well-ordering theorem," which threw the entire mathematical world—whether in support or in opposition—into a dither. He dealt with his adversaries in a 1908 treatise, the sarcasm of which is unparalleled in mathematical literature. In 1910 he finally received a full professorship at the cantonal University of Zurich. Shortly before the start of the World War he spent a night in the Bavarian Alps and when filling out the hotel registration form, in the space for "nationality" he entered: "Thank God not Swiss." By an unfortunate twist of fate, the head of the teaching department for the canton of Zurich stayed at the same hotel a short time later and saw the entry. Consequently, Zermelo's days at the University of Zurich were numbered. He was sent into retirement in 1916 and he resettled in Germany, where I had frequent scientific contact with him, even after I moved to Jerusalem. When I casually asked Zermelo's friend Erhard Schmidt why Zermelo virtually stopped publishing, he responded that Zermelo felt he could no longer expect to antagonize anyone with his publications.

It is thanks especially to this article that an important axiomatic system of set theory has carried the name "Zermelo-Fraenkel set theory" for three or four decades already. There are two reasons why my good fortune was undeserved. First, the full significance of my new axioms was not immediately recognized and exploited by me, but instead—as previously mentioned—a year later by John von Neumann. Second, this axiom and other ideas I presented in 1921, as well as some totally new, essential discoveries, were described in 1922, precisely a year later and totally independent of me, in Helsinki in a lecture by Thoralf Skolem, an extremely important Norwegian researcher who at the time was still virtually unknown. He published a paper based on his lecture in 1923 and was just as entitled to take credit for the axiom as I was. It is not uncommon in mathematics that scientists in various countries make discoveries almost simultaneously and yet totally independently of each other.

Skolem visited me a few years later for an extended period of time in Marburg, where he was also interested in Kurt Hensel's research in a different area. I remained in close contact with Skolem, who was extremely modest and hardly aware of his significance, until his death in 1962. I also visited him in 1954 at his home in Oslo. Fortunately, I was able to be of great service to him. An international congress of mathematicians took place in Oslo in 1936. In October 1937 I received a written inquiry from the dean of the School of Mathematics and Natural Sciences at the University of Oslo, asking which of two named candidates I could recommend for the vacant mathematics chair there. One was Skolem, who at the time had been working at a research institute in Bergen, where he had been given a position in 1929 on my recommendation. I sent off a comprehensive report to Oslo, which led to his receiving the appointment.

Whereas my article in the *Mathematische Annalen* was relatively easy for me, the mental exertion with which I accomplished the second major achievement of my scientific career, also in 1921, was great and planned. It dealt with the problem of the independence of the "axiom of choice" by Zermelo (1904) and Bertrand Russell (1906).[8] In this instance too I consulted Zermelo before my findings

[8]Thanks to the genius of Bertrand Russell, who for good reason was awarded the Nobel Prize for Literature, it is possible to offer even laypeople a notion of the meaning of the Axiom of Choice.

Imagine an infinite number of pairs of boots, assigned the numbers 1, 2, 3, …. If *S* is used to denote the set of boots contained in all pairs, then a subset of *S* can be constructively defined, which contains a single boot from each pair, for example the set of all left boots.

If, however, we have an infinite number of pairs of socks and if *T* is the set of all the socks, it is apparently not possible to constructively define a subset of *T* which contains a single sock from each pair, since the socks in each pair are, at least theoretically, identical, and we have no characteristic at our disposal that defines one of the two, such as we have with "left boot."

The Axiom of Choice asserts that despite this constructive impossibility, a set—and then not just one—does exist, which contains one sock from each pair. In general it claims that: If an infinite number of sets are given, all of which are nonempty and disjoint (i.e., containing no common elements), then there is (at least) one set that contains a single element from each of the given sets. The axiom is a concise example of a purely existential assertion that lacks any constructive content.

In 1921 I posed the question whether the Axiom of Choice is independent, just as Euclid's parallel postulate is independent of the other axioms of geometry, so that its negation leads to a

appeared in the *Berliner Akademieberichte* (Reports of the Berlin Academy [of Sciences]).[9]

I do not wish to say more about my various publications from 1921 to 1929 (or 1931), especially because many were reviewed in the third edition of my *Einleitung in die Mengenlehre* (1928). I shall mention only a more historical work, since the surrounding circumstances are not lacking in irony. Georg Cantor, the inventor of set theory, had died in January 1918. At that time the Berlin professor Ludwig Bieberbach—whom I mentioned earlier within the context of his being a proponent of a racial theory of mathematics in 1934—was the secretary of the German Mathematical Society (and by the way not your garden-variety anti-Semite). He wrote to me in December 1926, bothered that an obituary for Cantor still had not been published and "how difficult it was to arrange an obituary for such a truly eminent man. I therefore venture to ask if you would not feel disposed to take on this necrology."

I accepted, but since I was busy finishing up two books in 1927–1928, the 78-page Cantor biography did not appear until 1930 in the 39th volume of the annual report of the German Mathematical Society and also as an independent work, with the author reference: "A. Fraenkel in Kiel, currently in Jerusalem." I also published an abridged version in 1932 in the volume of the collected works of Cantor, edited by Zermelo, which was reprinted in 1962. This biography gave rise several years later to a futile attempt within Nazi circles to try to prove Cantor's "Aryan" descent. Cantor himself, however, according to some friends, had never made a secret of his Jewish heritage.

In 1929 the German Philosophical Society (DPG) asked me to be a special editor for an issue of their journal printed in gothic letters, *Blätter für Deutsche Philosophie*. This special issue would be dedicated to the philosophical foundations of mathematics. After initially declining I ultimately accepted their offer under

non-Euclidean geometry. In other words, it was about proving that the Axiom of Choice could not be derived from the other recognized principles of mathematics.

I was able to work out a proof in 1922, with the help of a "group theory" method that I invented. However, this proof was based on an assumption of set theory that was customary at the time, but to which I—and other scientists—had expressed certain objections. I generalized my proof in 1936, and starting in 1938 it was greatly advanced by the Polish-Jewish scientists Andrzej Mostowsky and Adolf Lindenbaum, as well as Elliott Mendelson and other Americans. But they were not able to exclude the assumption mentioned or a similar one. Finally, in 1963, it was Paul J. Cohen of Stanford University who managed in a masterful treatise to prove irrefutably the independence of the Axiom of Choice and of the Generalized Continuum Hypothesis.

[9] A. Fraenkel, "Der Begriff 'definit' und die Unabhängigkeit des Auswahlaxioms" ("The Notion of 'Definite' and the Independence of the Axiom of Choice"), *Sitzungsberichte der Preussischen Akademie der Wissenschaften, physikalisch-mathematische Klasse* (1922), 253–257. The Berliner Akademieberichte were officially called the Sitzungsberichte der Preussischen Akademie der Wissenschaften zu Berlin (Session Reports of the Prussian Academy of Sciences in Berlin). The article was published in the "Physics and Mathematics" section.**

the condition that I would be at liberty to select the topics and my colleagues, and that this would be expressly stated on the cover of the journal, mentioning my residence in Jerusalem. The double issue appeared in late 1930 and included significant essays by Heinrich Scholz, Rudolf Carnap, Karl Menger, Paul Bernays, Walter Dubislav, and myself. The publishers certainly regretted that a few years later!

My professional career began in March 1919 with, in addition to the lecture classes, an invitation to Göttingen to Privy Counselor Felix Klein, almost 70 years of age, but still active as the "foreign minister" of German mathematics. He had asked me to write about the number concept and algebra for the series he published, *Materialen für eine Wissenschaftliche Biographie von Gauss* (Materials for a Scientific Biography of Gauss). I wrote issue no. 8 of the *Materialien,* which appeared in 1920. Around that time I heard my first lecture by David Hilbert and during my visits to Göttingen in the subsequent years I met with almost all the leading mathematicians. I went on longer outings with Emmy Noether, who later gave me the great honor of giving a lecture on my explanation of the foundations of set theory.

In Marburg, and later in Kiel, I had an average of 30 to 100 students in my lecture classes, depending on the subject; that is, fewer than after the Second World War in Jerusalem. In my seminar classes there were usually between 20 and 30 students. My lectures covered the most varied subject areas of arithmetic and analysis, as well as individual branches of geometry. Based on the evaluation of critical observers, my lectures were too difficult at first but after a few years I adapted myself to the students who did not learn as quickly. A popular lecture class that I offered occasionally was on the "essence of mathematics," which also attracted my non-math students. During those years I had two doctoral candidates, one of whom later became the rabbi of the Darmstadt secession congregation, Julius Merzbach. The level of the mathematics students in Marburg corresponded to the average in German universities, but was far below what I later encountered in Jerusalem, especially with respect to the exceptionally gifted. Not until I was in Jerusalem did I have the chance to work with the four most outstanding students who made an international name for themselves in my area of research (Yehoshua Bar-Hillel, Azriel Levy, Michael Oser Rabin, Abraham Robinson). And in Germany I never experienced the rare luck of discovering a 12-year-old *wunderkind*, which did happen to me in Israel. Getting along with the students came naturally to me, even though nationalism and anti-Semitism were already widespread among the Marburg students—as at other German universities as well—and I remained resolute, for example, in the student dispute in 1925–26 over Theodor Lessing (who was murdered by Nazis in Marienbad in 1933).

For a brief time (1921–1922) my colleague as a *Privatdozent* was the highly talented young Helmut Hasse, who had also had Hensel as a doctoral advisor. My personal experience with him was only positive and his character was always irreproachable. When from Jerusalem I offered to resign from the editorial board of *Crelle's Journal*, which he headed, in order to avoid political difficulties, he indignantly refused to accept it. Some years later, after he had already become a professor in Göttingen, his life was shattered by a crisis: One of his adversaries discovered that he had a Jewish great-grandfather. Although the German race laws only went back to grandparents, and on top of that from his appearance and attitude he gave a completely "Aryan" impression, he saw his situation as intolerable. He turned to Hitler, who declared him—as he also did for other outstanding but not purely Aryan scholars—an honorary "full Aryan." Subsequent to that he joined the Nazi party, but did not seek any alibi after the war was lost, in contrast to the majority of career opportunists who changed their loyalties with the tides. When in June 1946 I met with one of the most eminent British mathematicians, Godfrey Harold Hardy, in Cambridge, I was shocked to hear these details about Hasse; Hardy was writing a letter to the British occupation authority in Göttingen, calling for Hasse, who had been dismissed from the university due to his membership in the Nazi party, to be reinstated due to his scientific importance. Hasse later received a professorship in Hamburg.

In early 1922 I was named a non-tenured associate professor (*Extraordinarius*) by the Minister of Education, which was an honor considering I received the position so early on; however, it brought with it no rights or salary. With respect to the latter, the teaching assignment I received in April 1922 was more significant. For readers who are not familiar with or who have forgotten about the hyperinflation from 1921 to 1923, it should be noted that a decree of September 12, 1923, "raised" the fees paid for a teaching assignment to 227.2 million marks per month (dollar exchange rate in 1921: 75 marks; early 1923: 18,000 marks; August 1, 1923: 1 million marks). I received unexpected recognition when the roughly 30 *Privatdozenten* and non-tenured professors in the School of Arts and Humanities (*Philosophisches Fakultät*) chose me to be one of their representatives. There were two or three such representatives who had a seat and vote in the sessions of the *Fakultät* board, which was all-powerful in academic matters.

In 1926 a man turned to me with a written, scientific inquiry, who several years later—despite his French name—would play a fateful role in the nazification of the Prussian universities. His name was Erhard Tornier and in the area of mathematics he worked closely with a very young and highly gifted mathematician from Yugoslavia, Willy Feller. Shortly after I received my full professorship in Kiel, I facilitated Feller's being able to do his own postdoctoral research in Kiel. In late 1931 Tornier had asked if he too could transfer to Kiel for his postdoctoral work, in order "to be able to be near me and learn a lot from me." I knew Tornier only superficially, but foolishly granted him his wish, particularly in consideration of Feller.

After the Nazis took power, in late winter 1933, it turned out that Tornier had long been a Nazi and, in view of the Nazi victory that he anticipated, had made all the necessary preparations for the mathematics chairs at all Prussian universities to be placed in "reliable" hands. Among his first steps was to determine Feller's "non-Aryan" background and drive him from the university. Feller went to Sweden and initially had to weather hard times, until he ultimately received a professorship commensurate with his importance at Princeton University, in the United States. Tornier also attempted to block Hasse, who had received a position in Göttingen, from entering the Mathematical Institute there. As I was told in a letter, he wanted to inherit my chair in Kiel. Because I had left Germany and had resigned in writing from the Prussian public service, it wasn't really of concern to me. Nevertheless he wrote to me from Kiel in April 1933: "I emphasize everywhere how gentlemanly your conduct was. In my view you have done a service to Judaism."

When we visited the Land of Israel in March 1926, I had justified hopes of soon being able to move into a tenured university position. A non-tenured associate mathematics professorship at the University of Halle had opened up and of the three candidates for the position I was recommended to the Ministry of Education at the top of the list—thanks especially to the recommendation of Hasse, who had recently become a professor there. With that I was virtually assured of getting the position. When we returned, however, to my dismay it turned out that the Ministry had withdrawn the position from the mathematics department and wanted to allocate it instead to applied physics. I do not need to emphasize that the years 1926–1927 were filled with professional disappointment and financial concerns.

But decisive changes came to pass. I have already mentioned (Chap. 3) that Otto Toeplitz left Kiel to assume a professorship in Bonn in the fall of 1927, which gave me new hope. Winter progressed with a lot of worries. As favorable as the list of candidates was for me, the talented *Privatdozent* who was third on the list had influential supporters. When in March 1928, after visiting my mother, I spent a few days in the Bavarian Alps, part of the time accompanied by my friend Rudolf Carnap, an outstanding logician (at the time he was still in Vienna and Prague; later in Chicago and Los Angeles), I received a telephone call after the end of the Sabbath from my wife, who informed me of the letter from the Ministry, which contained congratulations on my receiving the mathematics chair in Kiel and an invitation to Berlin for negotiations. And so, in April I was delighted to embark on my post as professor and head of the Mathematical Institute at the University of Kiel, working together with my older colleague Ernst Steinitz. My wife had to remain in Marburg at first, where our younger son was born in mid-May.

Fraenkel's home in Kiel

Even before my wife was able to join me, however, I received a telegram in Kiel in early June from London. It had been sent by Dr. Judah Leon Magnes, chancellor of the Hebrew University, informing me that the university's Board of Governors was offering me the mathematics chair that was freed up when Edmund Landau[10]

[10]Although classes in mathematics at the Hebrew University were offered earlier (see reference to Dr. Amira on page 139), advanced study began in 1927 with the arrival of the celebrated mathematician Edmund Landau. Born in Berlin to an assimilated Jewish family related to the famous rabbi Yechezkel ben Yehuda Landau, author of the halachic work *Noda Biyhudah* (נודע ביהודה), Landau taught mathematics at the University of Berlin (1899–1909) and at the University of Göttingen thereafter.

left, at the same time making me the director of the newly opened Institute of Mathematics. In some respects it was lucky for me that I received this position in Jerusalem after having received the professorship in Kiel and not before. Although many German Jews found my decision incomprehensible, no one could accuse me of having been wooed to Jerusalem due to external circumstances.

Edmund Landau

While I was extremely happy about the professorship, I did not conceal the complexity of the situation. I had just taken on the chair in Kiel and my wife was still in Marburg with four small children. On top of that, however, my colleague Steinitz fell ill at the beginning of the summer with an incurable heart condition, making it impossible for him to return to his teaching position, and leading to his

He studied Hebrew as an adult and spoke at the groundbreaking opening gala ceremony of the Hebrew University in 1925. In 1927 Landau moved to Palestine/Land of Israel with his family and taught at the Hebrew University and headed the mathematics department and the Institute of Mathematics, but decided to return to Germany in 1928, finding the power struggles at the Hebrew University intolerable. Prior to leaving, he wrote a letter to the university authorities proposing his successor in heading the department. Fraenkel was his first choice, and Landau suggested he be appointed full professor as well as head of the Institute of Mathematics. Michael Fekete from the University of Budapest was Landau's second choice, whom he proposed be invited as a lecturer. The university invited Fraenkel in 1928, after he had just accepted a professorship at Kiel University. Fraenkel, intent on eventually taking the position at the Hebrew University, requested a two-year leave of absence from Kiel in 1929. Meanwhile, the Hebrew University approached Fekete, who immediately accepted. Fekete assumed he would be appointed full professor and head of the Institute of Mathematics, since Fraenkel had not responded positively. Shortly thereafter Kiel approved Fraenkel's request for a leave of absence and Fraenkel announced he would accept a position in Jerusalem as a full professor and head of the Institute for a trial period of two years. An embarrassing dilemma ensued. Fraenkel had an advantage of prestige and rank in a university abroad, yet Fekete had already been promised the job. The resulting compromise was that Fraenkel be appointed full professor, while Fekete received a rank of professor ad personam, that is, he was assigned a "personal chair" established specifically for him, but his formal position and salary were those of a lecturer. The institute was to be codirected by the two of them, simultaneously or sequentially. Fekete began teaching in Jerusalem in late 1928, whereas Fraenkel first started in the fall of 1929. See Katz, S. "'Pure Science' in a National University: The Einstein Institute of Mathematics and Other Research Institutes at the Hebrew University during its Formative Years," in Shaul Katz and Michael Heyd (eds.), *The History of the Hebrew University of Jerusalem: Origins and Beginnings* (in Hebrew) (Jerusalem: Magnes Press, 1997), 418–419.**

death in late September. I therefore could not assume the position in Jerusalem in the fall of 1928. As the only full professor I was responsible for finding a successor for Steinitz. It was easier for me to assume responsibility for delaying the start of my new position until fall 1929, since the Board of Governors had decided at its June 1928 session to offer a second teaching position for mathematics, in addition to mine, to Dr. Michael (Mihály) Fekete of Budapest.

Dr. Michael (Mihály) Fekete

On the other hand, whereas for the winter semester 1928–1929 there were only two math instructors at the University of Kiel, the *Privatdozent* Robert Schmidt and myself, and that was not sufficient, I found a solution that greatly satisfied both the Ministry's and my needs. I requested that Jacob Levitsky, a Russian living in Palestine/Land of Israel, who had completed his doctorate in Göttingen with Emmy Noether as his advisor, be hired as an assistant in Kiel, to ease my situation until someone could be found for Steinitz's chair. That way I had a chance to practice my conversational Hebrew during the winter while at the same time enjoying scientific stimulation. Levitsky later became a professor at the Hebrew University and one of the leading algebraists of his time.

The years 1928 and 1929 were interesting for my career, although extremely demanding in other respects. In summer 1928, in addition to complications involved with relocating my family to Kiel due to the housing shortage, I was supposed to complete the corrections to the greatly expanded third edition of my *Einleitung in die Mengenlehre* (Foundations of Set Theory). In September I also had to represent the University of Kiel by giving a talk at the International Congress of Mathematicians in Bologna, a congress that Luitzen Egbertus Jan (L. E. J.) Brouwer of Holland passionately opposed.[11] Furthermore, I assumed other

[11]From a review of W. P. van Stigt, *Brouwer's Intuitionism*, Studies in the History and Philosophy of Mathematics, vol. 2 (North-Holland, 1990), by Wim Ruitenburg in: *Modern Logic* 2 no. 2 (July 1992), 428: "In 1928 the situation reached a climax when Hilbert headed a German delegation to the International Congress of Mathematicians at Bologna, and Brouwer supported a German boycott because of the way Germans had been discriminated against at these meetings since the end of World War I."**

obligations of this kind in the fall of 1929, after we had already left Kiel and spent the month of August in a chalet we had already rented in 1928 in the beautiful village of Celerina. In September I gave lectures at congresses in Prague, Salzburg, Vienna, and Warsaw. In Vienna, the distinguished logician Moritz Schlick—who, although he was not Jewish, was shot at the university in June 1936 by a Nazi student—kept the philosophy department open on Sunday, where I was then able to carry on an extensive, incredibly inspiring discussion with the founder of the Vienna Circle. Most interesting by far was Warsaw, where I had been invited by the Polish government and the Congrès des Mathematiciens des Pays Slaves[12] to give one of the eight plenary lectures. The hospitality of the Poles—even those who belonged to anti-Semitic parties and generally required proof of baptism before issuing a professorial qualification—was extremely cordial. Although I explained to the president of the Congress, who invited me and others to lunch, that I lived an orthodox ritual lifestyle and ate my meals with the Jewish Sejm[13] representative Joshua H. Farbstein, the well-known head of the Mizrachi movement and president of the Jewish community of Warsaw, the warmth of our relations continued undiminished.

It was remarkable how the Kiel university authorities reacted to my receiving a position in Jerusalem. The curator (that is, the Ministry's administrative representative at the university), whom of course I duly notified about the position in Jerusalem, made a snide remark. I had the impression that he was happy, possibly to have gotten rid of a Jewish professor, and a naïve fool at that, who took seriously the notion of trading a Prussian professorship for such an exotic and insecure position in the Middle East. The then rector of the university, the expert in international law Walter Jellinek, had the opposite reaction. He had officially welcomed me at the June session of the University Senate as the university's new full professor, but at the same time emphasized the honor of my having in the meantime received a professorial appointment at the Hebrew University. A subtle undertone resonated along with this welcome to the extent that Jellinek—son of a baptized (but formerly Jewish) father and a non-Jewish mother, who had himself married a Christian woman—although hardly aware of his own Jewish background, was actually the grandson of Adolph Jellinek, the well-known chief rabbi and scholar in Vienna. Walter Jellinek's father Georg Jellinek had been considered one of the greatest authorities on constitutional law, most recently working as a professor at the University of Heidelberg, where Walter was also offered a professorship while in Kiel. Despite his name he was never perceived as Jewish. Nevertheless, he of course lost his professorship when the Nazis took power.

My wife and children spent the holidays in the fall of 1929 in Amsterdam, and I was at my mother's in Munich. Directly afterward, we met in Arth-Goldau on the train to Milan, in order to travel via Genoa–Alexandria to Jerusalem, where in early November 1929 I took up my professorship at the Hebrew University—initially

[12]Congress of Mathematicians from the Slavic Countries.**

[13]The Sejm is the lower house of the Polish parliament.**

only for two years. I still hold that position (on leave from late 1931 until 1933), emeritus since the end of 1959.

My report about this last phase of my life in Germany should not close without my describing some people who in every respect deserve to be highlighted. Those who first come to mind are eight scientists. Of course I cannot and do not wish to offer biographies or acknowledgments of their scientific accomplishments that can be easily found elsewhere. Instead I will mention primarily those aspects that were significant for my own development. Of these eight men, there are four mathematicians: Hilbert, Brouwer, Landau, and von Neumann; two physicists: Einstein and Niels Bohr; and two Protestant theologians and philosophers: Rudolf Otto and Heinrich Scholz.

David Hilbert

In his time David Hilbert (1862–1943) was the most significant mathematician in the world. For a long time he shared this honor with Henri Poincaré, who died in 1912. In contrast to most of his colleagues, Hilbert's discoveries in successive periods encompassed the broadest range of pure mathematics. He hardly dealt with applied mathematics, except for one not very successful period devoted to physics. He was born in Königsberg and never relinquished his East Prussian accent. The number of true anecdotes about him is legion, as he was without doubt a highly original character. He became a professor in Göttingen in 1895 and declined appointments to Leipzig, Berlin, Heidelberg, and in 1919 to Bern. He was correctly considered the scientific head of German mathematics, and was acknowledged throughout the world. Students flocked to him from all over Europe and the United States. At the second International Congress of Mathematicians in Paris in 1900 he gave a programmatic lecture on "Mathematical Problems."[14] The 23 key unsolved problems he enumerated largely determined the developments in mathematics in

[14]See David Hilbert, "Mathematical Problems" (Paris, 1900), trans. Mary Winston Newsom, in *Bulletin of the American Mathematical Society* 8 (1902), 437–479.**

the subsequent decades. Most of these problems have since been solved,[15] problem no. 1 by Paul J. Cohen in 1963.

Hilbert always remained free of all national and racist prejudices. After the turn of the twentieth century he had a large number of Jewish students, both in absolute and relative terms. In his own working life he was greatly influenced by two Jews, Adolf Hurwitz and Hermann Minkowski. His authority and tenacity managed twice to break through the prewar prejudice in the Prussian Ministry of Education and Cultural Affairs against appointing Jews to full professorships: Minkowski in 1902 and Landau in 1909. He pushed through the postdoctoral professorial qualification (*Habilitation*) granted Emmy Noether at the University of Göttingen despite the resistance of many colleagues, including some Jews, although this resistance was chiefly directed to her being a woman, rather than on account of her being a Jew and salon communist.

Hilbert's response to a question of Bernhard Rust, the Nazi Reich Minister for Science, Education, and Popular Culture, was typical. At a banquet in 1934 in Göttingen, Rust asked: "Is it really true, Mr. Professor, that your institute suffered so much from the departure of the Jews and their friends?" to which Hilbert replied, in his characteristic East Prussian dialect: "Suffered? No, it hasn't suffered, Mr. Minister. It simply doesn't exist anymore!"

I never studied in Göttingen so I was never actually a student of Hilbert. It is thanks to a fortuitous occurrence that I nevertheless had the opportunity to develop a close relationship with him. His last creative period, starting in 1918, was dedicated to axiomatics and the foundation of mathematics in general. Totally unrelated to that, in 1921, I had taken on the axiomatics of set theory, the fundamental discipline of mathematics, to which Hilbert had always attached great significance, and whose creator, Georg Cantor,[16] he had defended against harsh criticism. Hilbert's highly talented assistant and close collaborator Paul Bernays (descendant of the famous Hamburg Talmud scholar *Chacham*[17] [Isaac]

---

[15]See also "David Hilbert zur Feier seines 60. Geburtstages," in *Die Naturwissenschaften* 10, no. 4 (1922); Hermann Weyl, "David Hilbert and His Mathematical Work," *Bulletin of the American Mathematical Society* 50 (1944), 612–654.

[16]Georg Cantor (1845–1918), was of Jewish descent, but he was not related to Moritz Cantor, the Jewish historian of mathematics, although the two are often mistaken for one another. Georg Cantor was not only a great researcher, but in fact a genius, who had established a new mathematical discipline, *set theory*, within a short period of time (1873–1897), with completely innovative ideas and unrelenting tenacity despite the resistance of almost all his contemporaries. Set theory had already unfolded in the first decade of the twentieth century into a foundation upon which all of mathematics is built. Thus it became a tool that could be used to methodologically prove the basic mathematical concepts; on the other hand, it also became the connecting link between both mathematics and logic, and arithmetic and geometry.

The applications of set theory within the theory of functions, geometry, topology, probability calculus, and other areas are indisputable. Hilbert and Minkowski are among the few who, already in the early days of set theory, regarded Cantor as one of the most original researchers. The "intuitionists," however, continue to reject set theory even today. The vast majority of mathematicians and philosophers nevertheless agree with Hilbert's dictum that "No one shall expel us from the paradise that Cantor created for us."

[17]Chacham is a Hebrew title of respect for a wise man with extensive knowledge of Jewish texts.**

Bernays)—who taught at the Swiss Institute of Technology (ETH) in Zurich starting in 1933—focused as of 1935 specifically on set theory. As a result of my own engagement with set theory, Hilbert asked me in 1922 to give lectures at the Göttingen Mathematical Society. A relentless and yet objective critic on such occasions, he must have been impressed with my results, because he offered me warm praise. Our further communication took place mostly in writing. Although he had accepted my invitation to hold a lecture in Kiel, due to external circumstances this did not come to pass. After the third edition of my *Einleitung in die Mengenlehre* (*Foundations of Set Theory*) was published, I received a very cordial handwritten letter of thanks. Not until much later did I learn that he had considerable influence on my receiving the appointment in 1928 as a full professor in Kiel, since the Ministry had sought his opinion on the various candidates.

L. E. J. Brouwer [1881–1966] of Holland can be viewed as the decided adversary of Hilbert. His mathematical merits are limited to two—to be sure, very important—achievements. He mastered, especially from 1911 to 1914, certain difficult topographical problems that have to do with the concept of dimension; however, this was the virtual extent of his actual mathematical work. But even earlier, and also in the years 1918–1930, he had developed an original doctrine in an ultra-dogmatic manner that he referred to as "intuitionism," and later "neointuitionism," in which he rejected the greatest part of "classical" mathematics of the last three centuries as well as "classical" (Aristotelian) logic, with the argument that these were in part meaningless and in part incorrect teachings, since they were based in particular on "existence" rather than "construction."[18]

[18]The most well-known of the three principles of Aristotelian logic is the Principle of the Excluded Middle (*tertium non datur*), which Aristotle himself expressed approximately in the form of the following disjunction: In an entirety ("set") of objects, either every object has a given property *E*, or at least one object has the property arising from the negation *non-E*.

In 1907 Brouwer charged that in the "classical" mathematics of the last three centuries, most assertions—especially in infinitesimal calculus, the theory of functions, set theory, etc.—are either false or meaningless, because they utilize existential judgments instead of being restricted, as with the Greeks, to constructions.

According to Brouwer, Aristotle tacitly considered only finite sets, in which one could—as least in principle—examine every individual independently. And that is why the disjunction applies, Brouwer claimed, not because of the Aristotelian principle, but in anticipation of the individual examinations. He agreed, furthermore, as Hermann Weyl (who was not Jewish but nonetheless left Göttingen after the Nazis took power and lived in Princeton, New Jersey) declared in 1921, that:

> We must be careful about thinking of infinite sets not merely as defined by a property that is characteristic of its elements, but as a set whose elements are so to speak spread out in front of us, so that we can run through them just as an officer in the police office goes through his file to determine if an element of this or that kind exists within the set. With respect to an infinite set, this view of the mathematical universe is an attractive but rather unrealistic idealization. [cited in part from Dirk van Dalen, *Logic and Structure*, 5th ed. (London/Berlin: Springer, 2013), 155].

The actual mathematical conclusions from the neointuitionist standpoint, which in fact invalidate most of modern mathematics and almost all of set theory, cannot be elaborated on here. With respect to logic, however, scientists have concerned themselves very much since 1930 with the

His standpoint was rejected by the overwhelming majority of mathematicians and logicians. His discussions with Hilbert in particular assumed a personal note. The two virtually drowned one another with their cursing. The dispute took on a particular nuance in the sense that the Dutchman Brouwer set himself up as a champion of Aryan Germanness. Consequently Hilbert removed him from the editorial board of the *Mathematische Annalen* after he objected to what he felt were too many Eastern European Jewish ("*Ostjuden*") authors. In 1922 he published a 16-page Open Letter of protest to the Dutch Minister of Education for allowing the outstanding French scholar Arnaud Denjoy, who was professor at the University of Utrecht and a member of the Amsterdam Academy of Sciences, to maintain a "corresponding membership" although Denjoy conducted himself, according to Brouwer, like an "agent for the French government." Brouwer also openly protested against participating in the International Congress of Mathematicians in Bologna in 1928, claiming that the many German scholars, who under Hilbert's leadership participated in the Congress along with the French, had sinned against the "Manes[19] of Gauss and Riemann." Luckily, his instinct saved him—in contrast, mutatis mutandis, to another outstanding Dutch mathematician—from accepting a position at the University of Berlin, which the Nazis offered him in 1933.

I often got together with Brouwer, especially in Amsterdam, and was among the first mathematicians outside of Holland who treated his intuitionist views in lectures, articles, and books in a non-dogmatic, often critical form. At first Brouwer was enthusiastic in his praise of me, but I later became the object of his passionate and grievous rejection. My presentations of intuitionism were the earliest, but by no means the most in-depth. Starting in 1930 many esteemed mathematicians from Holland, the United States, and other countries distilled from Brouwer's ideas all that made sense and lacked dogmatic prejudice, and was insightful and sometimes groundbreaking also in classical mathematics.

Brouwer was temporarily suspended in 1945 by the Dutch government because of his connections to the Nazis, especially during the German occupation of Holland, when the Dutch Jews suffered extremely cruel treatment and most were exterminated in camps, but he later regained his position. I found it impossible to resume our former friendly relations.

Edmund Landau (1877–1938) can also be described as an original sort of character. Starting in the 1920s he took on the middle name "Yechezkel," after the Prague rabbi and Talmudist Yechezkel Landau, whose brother was an ancestor of his. Edmund Landau's father was the well-known Berlin gynecologist Professor Leopold Landau, who, after originally being rather distant to Judaism, later espoused the national-Jewish stance and supported the cause of the Palestine/land

logic of Brouwer, Arend Heyting, and Stephen Cole Kleene, and these investigations—to the extent that they distance themselves from a dogmatic rejection of classical (Aristotelian) logic—have brought forth much new and interesting material.

[19]Manes are spirits of the dead (as those revered as minor deities in ancient Roman religion).**

of Israel Jewish Colonization Association (PICA). He proudly showed me Hebrew inscriptions in the courtyard of his mansion near Berlin's Pariser Platz.

Landau was one of the most prolific scientists of his time. From 1904 to 1909 alone he published more than 50 papers, some of which were very extensive, and a two-volume, innovative book on the distribution of prime numbers. His main areas of research were the theory of functions and analytical number theory. He worked together with the outstanding English mathematicians G. H. Hardy and John Edensor Littlewood, with Harald Bohr, the brother of Niels Bohr, and many others. Over time he developed an increasingly concise, concentrated style of writing, not only in his papers but also in his books, totally erasing the intuitive dimension of his mathematical problem-solving—in a tradition first introduced by Gauss. Not everyone likes this style, especially when dealing with such difficult subjects. Apart from that, he did not lack a sense of humor, as demonstrated by his prefaces to *Foundations of Analysis*,[20] which was widely read in both German and English.

In 1899 Landau completed his doctorate[21] and in 1901, at 24, his postdoctoral professorial qualification, both at the University of Berlin. After the premature death of Minkowski, Hilbert wanted a first-class successor, from whom he anticipated creative talents. And so it happened that in 1909, Landau, at 32 years of age, received the important Göttingen professorial position, in every sense an unusual appointment, skipping the then usual intermediate step of non-tenured associate professor (*Extraordinarius*). He was very successful as a teacher in Göttingen and attracted doctoral candidates from all over the world. His relationship to Hilbert, however, turned out not to be as agreeable as had been expected. After Felix Klein (dubbed the "foreign minister of mathematics") retired, Hilbert appointed the Jewish mathematical analyst Richard Courant as his successor; Courant is credited especially for the establishment of Göttingen's prestigious Mathematical Institute after the First World War, with funding from the United States. In 1933 special efforts were made to try to keep him, but he was lucky to soon receive an academically and administratively more outstanding position at New York University.

Landau's avid wish to receive a professorship in Berlin did not come true. His quirky nature had made him many enemies, especially Ludwig Bieberbach, who had enjoyed a full professorship in Berlin since 1921. By virtue of his parents' and his brother-in-law Paul Ehrlich's assets, Landau was a wealthy man. He built a mansion-like home in Göttingen to house his huge scientific library. As a guest I often had the chance to admire his lavishly designed home. Aside from mathematical research, to which Landau was dedicated with untiring diligence, he had two hobbies: He owned an extensive and probably valuable postage-stamp collection, which he constantly expanded, and in his free time he read detective stories with a passion.

[20]*Foundations of Analysis*, trans. F. Steinhardt (New York: Chelsea, 1951).**

[21]It is characteristic—less of Landau than of the nature of mathematical research—that this seminal doctoral dissertation was only 12 pages long. The curriculum vitae was written in Latin.

After the founding of the Hebrew University in Jerusalem in 1925 with initially three institutes, it seemed natural that a chair should also be established for mathematics, a science often characterized as "Jewish." In addition to the two world-famous mathematicians Jacques Hadamard and Edmund Landau, the university's Board of Governors also included Selig Brodetsky who in 1924 had taken up a position as professor of applied mathematics at the University of Leeds and who as a Zionist leader had rendered great service to the university. At first Einstein was also a member, but he soon withdrew from the Board of Governors. Shortly after the 14th Zionist Congress in Vienna, the Board of Governors of the Hebrew University met in September 1925 in the home of Dr. Elias Straus in Munich. On the agenda was also the question of the appointment of a mathematics chair. As I remember, it was the last (or even the only) session in which Einstein participated. I was also in Munich at the time, at my father's deathbed, and Landau visited me. Obviously, there was a proposal that Landau go to Jerusalem either temporarily or long-term. He was in fact offered the position and was prepared to accept it. Although he had not been raised close to Jewish tradition, much less Hebrew scripture, he made an effort in his usual energetic fashion to learn Hebrew and managed amazingly quickly to master the language, especially to speak it fluently. And so he went to Jerusalem in the fall of 1927 as the first professor of mathematics. Supported by associate professor Dr. Benjamin Amira, who was raised in Palestine/Land of Israel and had completed his doctorate at the University of Geneva,[22] Landau successfully taught the eager students for the duration of the semester.

This is not the place to discuss the differences that arose between Landau, the university management, and the Board of Governors. In any case, Landau left Jerusalem at the end of the winter semester and never returned. Thus the Board, when it convened again in June 1928 in London, had once again to place the appointment of the mathematics chair on the agenda.

Edmund Landau had also paid attention to my prewar publications and followed my work from the early 1920s even more closely, although it was in a field that was relatively unrelated to his. In 1927, when I was still a non-tenured associate professor (*Extraordinarius*) in Marburg, he had already spoken with me somewhat tentatively of his intention to suggest me to the Board of Governors for the professorial chair in Jerusalem. I did not take him too seriously, especially when I heard that Hadamard wanted to suggest the distinguished Polish-French analyst Szolem Mandelbrojt; besides that, Landau's own decisions before he left for Jerusalem were anything but firm. Thus it was a great surprise for me when on June 5, 1928, in Kiel, only a few weeks after I assumed my position there, I received a telegram from London offering me the chair in Jerusalem. As Landau later explained to me, Mandelbrojt had decided not to leave France, so his candidacy had never even been considered. And he did in fact receive a position in Paris a

[22]Amira studied with Landau in Göttingen 1921–1924, then returned to Geneva, and later joined Landau in the Mathematics Institute of the Hebrew University.**

short time later, as Hadamard's successor at the Collège de France. My appointment had been unanimous, with even the approval of the physicist on the Board of Governors, Leonard S. Ornstein from the Netherlands.

Landau's and my relationship became increasingly close in those years, encouraged by his growing competence in Hebrew. He even sent me telegrams in Hebrew from Göttingen to Marburg![23] My last encounter with Landau was in Switzerland shortly after the Nazis came to power. In contrast to Courant, he was not frightened, but rather filled with an incomprehensible optimism. But the blow came soon—even sooner for him than for other Jewish professors. The Göttingen students demonstrated against him—evidently having been incited from higher up—and demanded his dismissal. A certain lecture sheds light on the background to this "resolute rejection" of Landau. It was held by Berlin professor Ludwig Bieberbach in spring 1934 in Berlin at the annual conference of the *Verein zur Förderung des Mathemathischen und Naturwissenschaftlichen Unterrichts*.[24] The press, for example the *Deutsche Zukunft* (German Future) of April 8, 1934, and *Forschungen und Fortschritte* (Research and Progress) of June 20, 1934, gave extensive coverage of the lecture. Cited here are some passages from it:

> The conduct of the Göttingen students toward Edmund Landau is well-founded and justified, as the Landau case clearly shows that there is a German and a Jewish mathematics, two worlds, separated by an unbridgeable chasm. The selection of the problems and the style of treating them is characteristic of the thinker and therefore a product of his racial affiliation. ... A people that has come into its own cannot tolerate such teachers and must reject foreign thought.

Bieberbach then made reference to the psychological typology of Erich Jaensch, assigning the abstracting *S* type to the Jews and speakers of Romance languages, and that of the *J* type, which seeks to understand reality in all its manifold aspects, to the Germans. He remarked on the explication of imaginary numbers by Cauchy and Goursat (both of whom are not Jews!):

> The Cauchy-Goursat theorem arouses intolerable displeasure in us Germans. But certainly not in a Landau. Such a thing is mentally determined. ... All in all, such juggling with concepts and a pronounced shrewdness characterize the nonorganic, hostile *S* type, especially Jewish mathematics.

[23]The following was a typical experience: In August 1928 on the way to the Congress of Mathematicians in Bologna, I visited Landau in his hotel in St. Moritz, where we conversed entirely in Hebrew. When he was briefly called away to a telephone call, I spoke in the meantime to the hotel porter about something, and when Landau returned I absentmindedly resumed the conversation with Landau in German, which prompted him to call out to me, appalled: באיזו שפה אתה מדבר אלי?! (In what language are you speaking to me?!).

[24]Association for the Promotion of Instruction in Mathematics and Natural Sciences.**

This was then compared to the "Eastphalian" Gauss, the "Nordic-Dinaric" Felix Klein, and the "Eastern Baltic–Nordic" Hilbert:

> Abstract Jewish thinkers of the *S* type knew how to deform [the axiomatics of Hilbert], so that they can be used as intellectual varieté. ... This is a typical example of how race-alien influences and race-alien seduction blocks the Germans from the source of their own strength. ... There is no self-sufficient mathematical domain that is independent of ideology and life; the dispute on the fundamental principles that is now raging is in reality a dispute over race.[25]

The journal *Deutsche Zukunft* drew its own conclusions:

> For practical cultural policy it turns out that mathematics is freed from the curse of sterile intellectualism; its burden falls upon those thinkers who are alien to the nation and the race, who will no longer exist in the future and whose representatives belong to the past and can no longer be viewed as German scientists. German mathematics is rooted in blood and soil. ...

These statements caused incredible outrage abroad, perhaps more than did the brutal dismissal of Landau himself and all the Jewish and "half-Jewish" scholars in Germany. In a letter to Bieberbach of May 19, 1934, one of the most eminent American mathematicians, Professor Oswald Veblen of the Institute for Advanced Study in Princeton, New Jersey, expressed his disgust at the distinctions made between Jewish and German mathematics and at such pseudo-scientific justification of the measures taken against Landau, "one of the great mathematicians of our epoch." He added that the article in *Deutsche Zukunft* was received in the United States "with varying degrees of sorrow, derision, and contempt." Veblen had written this letter in an outburst of pure humanitarianism, without having been able at the time to foresee that the expulsion of all "non-Aryan" scholars from Germany, and later also from Austria, would make the hospitable United States (together with Russia) into the center of the world for mathematics and physics, even before the end of the 1930s, particularly in view of modern nuclear physics. The consequences of this shift are generally known.

After losing his professorial chair in Göttingen, Landau moved to Berlin. Although he also went abroad to give lectures, he never managed to decide to emigrate, which would have meant losing all his assets. He died in 1938 at the age of only 61.

---

[25]There is no need to emphasize how absurd this racial distinction is. Bieberbach of course enjoyed the applause of German and Austrian mathematicians such as Wilhelm Blaschke and Theodor Vahlen, who, after being dismissed from his professorship in Greifswald in 1927, was appointed a full professor in Berlin in 1934, during the Third Reich. From 1937 on there was even a new professional journal *Deutsche Mathematik*.

John von Neumann

The most outstanding and at the same time versatile mathematician in the world in the second quarter of the twentieth century was Johann (later John) von Neumann (1903–1957). Far beyond the boundaries of mathematical science, where among many other things he developed the Theory of Games, which is so important today, he mastered the most difficult problems in quantum mechanics and other areas of theoretical physics, and in the areas of economics, meteorology, and above all in the theory and practice of electronic computers. His achievements and his almost prophetic foresight allowed him to play a leading role in the production of the atom bomb (starting in 1943) and the MANIAC[26] (starting in 1951), the machine that enabled production of the hydrogen bomb. Even as a young man he had developed a passion for inventing mechanical toys that—based on the tales of E. T. A. Hoffmann—earned him the nickname "Dr. Miracle."[27] However, hardly a year after he was appointed a member of the U. S. Atomic Energy Commission (1954) by then President Dwight Eisenhower, symptoms of a fatal illness were discovered, which led to great suffering and his death two years later.

Of course this is not meant to be a description of von Neumann's life's work,[28] but I would like to say something about our relationship, which however was limited primarily to the 1920s. Johann von Neumann, born in late December 1903 in Budapest, was the son of Max Neumann, a wealthy Jewish banker who

---

[26]This acronym was thought up by J. von Neumann and stands for Mathematical Analyzer, Numerical Integrator, and Computer.

[27]See Robert Jungk, *Brighter than a Thousand Suns: A Personal History of the Atomic Scientists*, trans. James Cleugh (New York: Harcourt, 1958) [German: 1956], 298: "Even during his years as an undergraduate at Göttingen, the ingenious Hungarian von Neumann had been nicknamed 'Dr. Miracle' by his fellow students, because of his passion for mechanical toys. They were thinking of [the] weird builder of automata, who invented the life-size and lifelike doll Olympia and then fell desperately in love with it," from E. T. A. Hoffmann's stories *The Sandman* and *Councilor Krespel*.**

[28]For that see S. Ulam, "John von Neumann," in *Bulletin of the American Mathematical Society* 64, no. 3 (May 1958): 1–49.

had been granted nobility status in 1913 by Emperor Franz Joseph. Throughout his lifetime, Johann (John) could never forget the communist regime of terror of Béla Kun in Hungary. He was recognized early on in school as a mathematical child prodigy. He received private tutoring and before he even completed secondary school it was clear that he would study mathematics. His father wanted him to learn something more practical, however, such as chemistry, rather than such an "unprofitable art" as mathematics. Consequently, von Neumann lived a threefold existence starting in 1921: he was formally registered as a math student at the University of Budapest, where—without actually studying there—he completed his exams each year and received his doctorate in 1925. He simultaneously studied chemistry at the Swiss Federal Institute of Technology (ETH) in Zurich and, also in 1925, passed his examinations for a graduate diploma in engineering. Mostly, however, he conducted mathematical research in Zurich and Berlin, either alone or together with professors there.

In 1923, Erhard Schmidt, who had been a professor in Berlin since 1920 and was one of the editors of the journal *Mathematische Zeitschrift*, sent me a lengthy manuscript while I was in Marburg entitled "Die Axiomatisierung der Mengenlehre" (The Axiomatization of Set Theory), written by Johann von Neumann, a name as yet unfamiliar to me. Schmidt, who a year earlier had presented a paper of mine to the Prussian Academy of Sciences, accompanied the manuscript with a letter saying that they could find no one in Berlin who was competent to evaluate the manuscript, and so he requested a statement from me whether a publication in the *Mathematische Zeitschrift* should be considered. The article, 84 printed pages, did in fact appear five years later in the 27th volume of the journal, but had already been published in essentially the same form in 1925 in Hungarian as von Neumann's doctoral dissertation in Budapest. At around the same time, with an accompanying letter of August 14, 1923, von Neumann sent me his work directly, and expanded it two months later with comments in a letter comprising 14 large quarto-sized pages.

It took considerable effort for me to work through the treatise, which not only deviated from everything that had thus far been published on the axiomatization of set theory, but it also contradicted previous ideas to the extent that it treated not sets but functions, and expressly permitted the "vast" areas that had been taboo for all predecessors. Far from understanding all that I read, I recognized from the small part I understood that the author must have a particularly extraordinary gift. To supplement our correspondence I therefore invited him to visit me in Marburg. His appearance made a huge impression not only on me, but also on my wife: the slim, not yet 20-year-old young man looked exactly as one would imagine a young genius.

During the short time that he stayed with us, I made it clear to him that it would be hard to expect mathematicians all over the world to digest such radical new ideas all at once. I therefore suggested that he first pick out the raisins from the cake, so to speak, as a program, the particular explication of which could be his article. He agreed, and as a co-editor of *Crelle's Journal* I arranged that the "raisin" article be published in volume 154 of the journal (1925). It would go beyond the present scope of this book to present the triumphal march of his ideas, which were later expressed in a thoroughly modified form by Bernays and Kurt Gödel.

That same year, a completely different treatment by von Neumann appeared in a Hungarian journal, yet written in German. It was a short text within the usual axiomatic framework titled "Zur Einführung der transfiniten Zahlen" (Introduction to Transfinite Numbers). Except for a small algebraic article published in 1922, this was his first publication, and it immediately proved to be a masterpiece. "Classical" to a greater extent than his doctoral dissertation, it offered not only the first rigorous definition of transfinite ordinal numbers, which had been introduced by Cantor in 1883 and widely used since then, but it was also the first foundation of a method that had been used without hesitation in many areas of mathematics because it had (erroneously) been considered intuitively obvious: namely "transfinite induction." For me, this treatment contained a special surprise. In 1921 I had given a talk, published in 1922, on Hilbert's initiative in volume 86 of *Mathematische Annalen*. In it I had added a new axiom, the "axiom of replacement," to the familiar axioms of set theory. The significance of this axiom now showed itself in a completely unexpected light, as von Neumann based the theory of transfinite numbers on my axiom in such a way that it proved it was absolutely essential for this purpose.

From 1927 to 1930 von Neumann was a *Privatdozent* in Göttingen and later in Hamburg. During this time we collaborated one last time, a collaboration that would have to be characterized as human rather than scientific. Assessing his situation rationally, he pointed out that in the coming years, fewer than half a dozen mathematics chairs would open up in Germany, whereas there were more than 30 *Privatdozenten* and non-tenured associate professors (*Extraordinarius*) hoping to receive one of these positions. For this reason he had decided to emigrate to the United States. His logic was faulty insofar as he had underestimated his own significance, which certainly would have brought him a professorial appointment. But his decision was a favorable one. In 1930–1931 he received a professorship at the venerable Princeton University in New Jersey, and in 1933 at the newly founded Institute for Advanced Study. He was the youngest permanent member of this institute, which was at the cutting edge of all scientific research institutions in the Western world.

In 1946, 1949, and 1952, when I held lectures in the United States, I accepted von Neumann's invitations to visit him in Princeton. When I pressured him in 1946 to finally visit Jerusalem, he answered very seriously: "Yes, I will in about six months, when I finish the electronic computer that I have been working on for so long." Many years earlier, in December 1935, he wrote to me in Jerusalem that he had heard from Professor Ladislaus Farkas, the Jewish-Hungarian chemist at the Hebrew University, how interesting and attractive Jerusalem was and he ended the paragraph with the comment: "And I hope to visit sooner or later." He did not suspect that his work on the computer and its continuation on incomparably larger and more complex machines would ultimately extend to the very end of his life. And so his wish was not fulfilled.

My relationship to Albert Einstein (1879–1955) extended into two areas: scientific research, of course, and also the Hebrew University, which was very dear to his heart.

Our first contact was arranged by Professor Ernst Steinitz, in Kiel, who told me in September 1920 about his meeting with Einstein during the "Kieler Herbstwoche," a week of events in Kiel related to the arts and sciences. Einstein had told Steinitz that he liked my small book that was published in 1919, *Einleitung in die Mengenlehre*. After the publication in 1923 of the second edition, which was twice the length, he had his secretary inform me that he read the book "with the greatest interest." Our personal communication, other than when we spoke to each other, was mostly through handwritten correspondence; it commenced after I followed up on his invitation in the mid-1920s to visit him on a trip to Berlin, and we engaged in a long, stimulating discussion.

I met with Einstein in Berlin, Brussels, and London until 1933, and in the years 1946, 1949, and 1952 at his home on Mercer Street in Princeton for many extensive conversations. Contrary to widespread belief, he was a professor in Zurich and Prague only for a short part of his long life, namely from 1909 to 1914. The greatest discoveries that have immortalized his name—both in the theory of relativity, which represented the end of Newtonian mechanics, and in molecular physics and the quantum theory, which marked the beginning of new radical developments in physics—are from the period from 1902 to 1909, when he was a technical expert at the Swiss patent office. In 1914, especially on Planck's recommendation, he was called to Berlin to be a full-time member of the Prussian Academy of Sciences and director of the Kaiser Wilhelm Institute for physics. In this capacity he was authorized to lecture at the university. But he remained a citizen of his chosen Swiss home. After he vacated these positions in March 1933, he worked until his death at the Institute for Advanced Study in Princeton—the final years he had emeritus status[29]—together with James Alexander, Kurt Gödel, John von Neumann, Oswald Veblen, Hermann Weyl, and outstanding physicists including, later on, Robert Oppenheimer. He did not often exploit the opportunity to give lectures, not least because his English was poor.

Since Einstein was a physicist and I was a mathematician, one cannot speak of direct scientific contact between us. As a student in Zurich he "by chance" had the great luck to hear a lecture on Riemannian geometry,[30] without which he could hardly have formulated his general theory of relativity in 1914–1916. In later years he always had mathematical assistants, among them two of my former students in Jerusalem, with whom I later reconnected in Princeton: Ernst Strauss (the youngest son of Dr. Elias Straus of Munich, whom I have mentioned numerous times), who later became a professor at the University of California at Los Angeles,[31] and after him until Einstein's death, Bruria Kaufman (wife of Professor Zellig S. Harris of Philadelphia), who in 1960 became a professor at the Weizmann Institute of

[29]The great French physicist Paul Langevin once said: The Pope of Physics has moved, and the United States will now become the center of the natural sciences.

[30]Riemannian geometry is a branch of differential geometry named after Bernhard Riemann (1826–1866), a student of C. F. Gauss.**

[31]Ernst Strauss (1922–1983) remained at UCLA until his death.**

Science in Rehovot.[32] I had a conversation with Einstein that was actually about mathematics in 1952, when he requested information from me about neointuitionism; this doctrine enticed him because of its "atomic" character, in contrast to "continuous" classical mathematics.

I knew far too little about physics to get involved in physics conversations with Einstein. Nevertheless I was continually amazed at his lack of understanding in later years vis-à-vis the new developments in the area of physics, as represented by Niels Bohr, Werner Heisenberg, Max Born, and others. Perhaps some of this can be attributed to a desired incomprehension, but the tragedy was that his objections ran parallel precisely to those with which the theory of relativity was attacked in the 1920s.

We seldom talked about ideological, that is, religious issues. He supported an almost Spinoza-like pantheism, and only in this sense, in my opinion, are those who consider him a "religious" person correct.

As generally known, Einstein was the main target of the race theorists who, under the leadership of the German Nobel laureates Philipp Lenard and Johannes Stark,[33] explained the theory of relativity as a "Jewish world bluff" and characterized the "Aryans" who thought differently—including Arnold Sommerfeld, Heisenberg, and most of the significant physicists—as "having a Jewish mind." Heisenberg, who in 1932 received the Nobel Prize for Physics, dared in a 1934 lecture on the changes in the foundations of the natural sciences to highlight Albert Michelson (an American-Jewish researcher who received the Nobel Prize for Physics in 1907) and Einstein in particular, and praised the special theory of relativity by referring to it as "among the absolutely certain foundations of physics today." Consequently, a certain senior high school teacher, *Oberstudienrat* Dr. Rosskothen, wrote in November 1934 to the Nazi Party's chief ideologue, *Reichsleiter* Alfred Rosenberg, saying that even if "the contemptible Jew Einstein received the Nobel Prize from the Swedes, those traitors to their race," it is not worth mentioning, and he asserted that the concentration camp is therefore the right place for Heisenberg, who should be charged with "treason against the people and the race." Rosenberg's office responded that the Reichsleiter agreed but, "in view of foreign opinion," reprimanding Heisenberg would have to suffice.

The special case of Rudolf Tomaschek is worth mentioning in this context. As a student of Lenard in Heidelberg, he was sent by his teacher in the 1920s to the Jungfraujoch to experimentally disprove the theory of relativity. Because his experiments instead confirmed the theory, he fell out of grace and had to transfer to the University of Marburg, where he was my colleague for a short time. In the Third Reich Tomaschek became a physics professor at the Technical University in

---

[32] She was a professor there until 1971, and then until 1988 at the University of Haifa. Bruria Kaufman returned to the United States in 1982, and died in January 2010 in Haifa.**

[33] Stark was dismissed from the University of Würzburg in 1922 for non-academic conduct. In the Third Reich he became director of the German National Institute for Physics. Incidentally, Stark, Lenard, and their supporters connected the struggle against "Jewish" physics with one against the mathematization of physics.

Munich, where he declared that modern physics was an instrument of world Judaism to destroy Nordic science. True physics, he said, was a creation of the German spirit.

Although Einstein and I did not have very close contact in purely scientific terms, worries about the establishment and development of the Hebrew University proved to be a common interest which led to our intensive and long-term connection. Since this did not become more intensive until after 1929, I will limit myself here to several remarks.

Toward the end of the First World War, Kurt Blumenfeld[34] had won Einstein over to the Zionist idea. In 1921 Einstein accompanied Chaim Weizmann, president of the World Zionist Organization, to the United States, in order to help out with publicity for the *Keren Hayesod* (United Israel Appeal, UIA). From then on he also became active in preparations to set up the Hebrew University, which opened in 1925. In 1924, on his way to Japan, Einstein visited the Land of Israel/Palestine for the first and only time, taking advantage of the occasion to give a number of lectures. But he wrote to me in 1936 from Princeton that he sincerely hoped to see the Land of Israel/Palestine again.

The Institute of Mathematics at the Hebrew University of Jerusalem, which I headed as of 1929, was named after Einstein, as was the physics institute that opened later. For a certain period of time he was also an active member of the university's Board of Governors, but he later withdrew entirely as a result of fundamental differences with Dr. J. L. Magnes, the chancellor of the university. In spring 1929 he sent me a personal, handwritten letter expressing his pleasure at my decision to go to the university, at first for two years, and he asked me to give him a verbal report during the summer vacation in 1930, which I did during a meeting at the Solvay Conference in Brussels. The next two years were replete with an intense correspondence over the future of the university.

In the fateful year, 1933, Einstein's relationship to the Hebrew University went through a crisis. In February Einstein had completed his scientific obligations in the United States and had booked his return passage. At that time he received initial warnings from Germany about the danger to his security, among others from General Hans von Seeckt, the creator of the *Reichswehr*, the armed forces of the Weimar Republic. Einstein traveled to Belgium and spent the months of April and May in Le Coq near Ostende, where the Egyptologist and Orientalist Abraham S. Yahuda visited him. It was through this man's influence, which I tried in vain to contain through correspondence with both Einstein and Yahuda, that the former allowed himself be carried away to publicly attack the Hebrew University. Instead of the Hebrew University, Einstein suggested opening a refugee university in Lugano, which I forcefully protested in my letter of May 4 to him.

---

[34]See Kurt Blumenfeld, *Erlebte Judenfrage. Ein Vierteljahrhundert deutscher Zionismus,* ed. and with an introduction by Hans Tramer (Stuttgart: Deutsche Verlags-Anstalt, 1962), 126ff.

[Blumenfeld was the secretary of the German Federation of Zionists (ZVfD) from 1909 to 1911, secretary general of the World Zionist Organization 1911–1914, and ZVfD president 1924–1933.**]

Weizmann asked me to come from Amsterdam to London where the joint efforts of several scientists finally managed to come up with a compromise on this issue, as well as regarding other matters of the internal structure of the university. Einstein, who had moved to Oxford in May and met us in July in London, promised to discontinue his campaign against the university and to return to normal relations on condition that an objective board of inquiry be sent to Jerusalem to review his accusations. This took place and a fundamental reorganization followed in the summer of 1935. From then on, slowly but surely, Einstein adopted a positive attitude toward the university.

Niels Bohr

My relationship to Niels Bohr (1885–1962), physics Nobel laureate of 1922, who became generally well-known through his work on the production of the atom bomb during the Second World War, can be depicted rather briefly; for me our contact was not quantitatively, but rather qualitatively significant and impressive. I knew his younger brother too, the distinguished mathematician Harald Bohr, who also worked on the papers of the Mathematics Class at the Hebrew University of Jerusalem and died more than ten years earlier than Niels. Harald was a friend of Edmund Landau and became a leading critic of Bieberbach's discussion on race in mathematics, which he told me about at the International Congress of Mathematicians in 1936 in Oslo. I participated in the congress, interestingly enough as a representative of both the Hebrew University and the British Mandate government of Palestine/Land of Israel. The Bohrs were "half-Jews" and born in Copenhagen, where they also worked throughout their lives. Niels received extraordinary honors from the Danish government.

My first and most impressive encounter with Niels Bohr was in September 1929, shortly before I moved from Kiel to Jerusalem. Bohr had come over from Copenhagen for the Kiel week of arts and science events, the "Kieler Herbstwoche." In addition to his public lecture, he also agreed to speak, after dinner in our home, to a small number of invited mathematicians and physicists, including Walter Kossel, on present and future developments of the natural sciences. Never in my life was I so profoundly impressed by a scientific talk, and many of the guests felt likewise. It was as if a prophet were speaking, not a physicist. I am no longer sure if the point of departure was the Uncertainty Principle, indeterminism, or the Principle of Complementarity. In any case his remarks were by no means limited to inorganic natural sciences, but overlapped with biology and psychology. Some references can be found in Bohr's work *Atomic Theory and the Description of Nature,* which was first published in Danish in 1929 and then in English in 1934.

I met with Niels Bohr—sometimes in the company of his wife and son—several more times in Israel, where he held lectures. Once, on his family's suggestion, I took them to attend Friday evening services in Oriental and Yemenite synagogues in Jerusalem; the conversations on these occasions were not as profound. In any case, except for perhaps Sommerfeld, no physicist—including Einstein—ever painted such a powerful picture of physics and its impact on human thought as Niels Bohr did.

It is strange that among all the non-Jewish Germans I encountered throughout my life, it was in fact two Protestant theologians with whom I had the closest, also personal, contact. I am thinking here of Rudolf Otto and Heinrich Scholz. In this context I could also mention Romano Guardini, a Catholic philosopher of religion and, starting in 1948, professor in Munich, but my contact with him was far less frequent.

Rudolf Otto 1925

Rudolf Otto (1869–1937) and I had in particular three areas of interest in common: the Hebrew language, German politics, and not least the philosophy of religion. Our philological conversations and correspondence referred in part to Hebrew material that I gave him and which he wanted to translate for his book and essays, and in part to some hypotheses he postulated for interpreting concepts and words from the New Testament, for which he sought my advice about their use in Talmudic writings. In retrospect I have to ask myself why Otto came to me for that kind of information, which was often rather specific, rather than contact an expert. During the time that we were both working at the University of Marburg, the reason would be the convenience of our frequent meetings and the personal discussions that were thus possible; later Otto presumably had become accustomed to viewing me as an assiduous and reliable responder to his questions.

After the upheaval of 1918 Otto was for a time a democratic representative in the Prussian National Assembly. From 1913 to 1918 he had already belonged to the Prussian House of Representatives. Around 1920 we had repeated political conversations, whereby I did not make any secret of my leftist sentiments, even if I was not a member of the Social Democratic party. Walther Rathenau, the German Foreign Minister, had negotiated the Treaty of Rapallo between Germany and Russia and pioneered a planned economy that was informed by sophisticated goals. After Rathenau was assassinated in June 1922, Otto came to me in an excited state, wanting to know what would happen then. I responded harshly that the murder was the logical consequence of the pussyfooting politics of the Democratic Party, which ultimately carried the moral responsibility for what had happened.

A few days later my wife and I were again invited to Otto. I did not find it unusual as we got together often. How surprised and awkward I felt that the other guests there were the leaders of the local group of the Democratic Party of Marburg under the direction of the theologian Professor Martin Rade, editor of the weekly *Die Christliche Welt* (The Christian World). After a few words of welcome Otto turned to me and said, "Now tell the gentlemen what you have to criticize about the Democratic Party!" At first I tried to avoid the request, but then I expressed my opinion openly, adding that it was not my intention to meddle in German politics, since as a Jew it was not my business. And that was, in fact, my last direct encounter with German politics.

As far as I could judge, Otto was the most profound and innovative German religious thinker of his generation, Friedrich Heiler and other eminent theologians notwithstanding.[35] His magnum opus, and at the same time one of the earliest research papers on the phenomenology of religion, is the book *Das Heilige (The Idea of the Holy)*, which was first published in 1917, when Otto had just assumed the chair in Marburg (the 28th printing came out in 1948). As he told me, and also said publicly, he first confronted the concept of the "numinous"[36] when, before the First World War, he attended most of the *Yom Kippur* service in a Sephardic synagogue in North Africa. Later, too, the service of the synagogue greatly influenced the development of his thoughts. For example, he often stressed that the term *kadosh* [holy] from the Old Testament is the most adequate

[35]Otto was a sharp contrast to Rudolf Bultmann, a New Testament scholar, not so much for reasons of professional expertise, but because the latter was utterly influenced by the philosopher Martin Heidegger, who was also teaching in Marburg at the time. Heidegger seemed to be more of a prophet than a teacher to his students, and Otto viewed him as a "seducer of youth." As we know, Heidegger publicly pledged his support to Hitler in 1933; with that he was attacking his own teacher in a shameful way, as the phenomenologist Edmund Husserl was born Jewish.

[36]Numinous refers to the sensation of being in the presence of divinity.**

characterization of the numinous. He was very interested in the substance of the prayers and the spirit behind them. Otto was most attracted to the liturgy of the High Holidays, which is why he asked us for a mahzor (prayer book) for the New Year (*Rosh Hashanah*) and the Day of Atonement (*Yom Kippur*). Certain sections, such as "*Uvchen ten pachdecha*" ("And so, too, instill Your awe") and "*Melech elyon*" ("Most High King"), enthralled him so much that he translated them and published them in part in the later editions of *Das Heilige* (starting with the eighth or ninth edition) or in *Die Christliche Welt.*

His absolutely selfless personality deeply impressed both my wife and myself. The three of us often discussed religious questions together during our Marburg years. However, he firmly rejected Hermann Cohen's understanding of Judaism. He felt particularly close to mysticism, as shown in his essay "Östliche und westliche Mystik" [Eastern and Western Mysticism] in *Logos* in 1924 (vol. 13, no. 1), 1–30.

Professor Carl Heinrich Becker, a friend of Otto, was an Orientalist who after the First World War became state secretary and later Prussian Minister of Culture and Education. He asked Otto to be the ministerial director in his ministry, but Otto declined this influential position because he did not want to have to restrict his research and teaching activities and because his health was declining in the 1920s. A temporary improvement in his health allowed him to take an extended study tour in winter 1927–1928 to India, Egypt, and Palestine/Land of Israel. He wrote interesting letters to me: from Mysore, where he had been invited by the maharaja; from Cairo, where he had visited important synagogues and in particular joined the German-Jewish ophthalmologist and Islam scholar Dr. Max Meyerhof; and above all from Jerusalem, where he had spent the Easter and Passover holidays. He knew the Land of Israel/Palestine from earlier visits, but this time he met with friends of mine: Simcha Assaf, Hugo Bergmann, Leo Arie Mayer, and especially Gershom Scholem; unfortunately he was not able to take advantage of an invitation from Dr. Magnes to hold a lecture at the university.

From around the mid-1920s until the Nazis took power and even later, Otto concentrated all his energy on establishing a museum for comparative religious studies, for which Marburg's university curator had already arranged appropriate space in the castle overlooking the city. In addition to precious objects from India, Egypt, Russia, China, and other countries, Jewish ritual objects were supposed to be included most prominently, in part promised through my referrals especially from Germany, Holland, and England. In late 1932 Otto was delighted to inform me that Rabbi Leo Baeck had presented the prospects of giving him "an entire synagogue."[37]

Although I never heard anything about the fate of the collection after 1935, the year of the last letter I received from Professor Otto, one can hardly assume that Jewish ritual objects in particular escaped confiscation by the Nazis. Otto's poor

[37]Following the Second World War and his liberation from Theresienstadt, Baeck visited me in Jerusalem following the official welcome I gave him in my position as Grand President of the Israel District of the B'nai Brith lodge.

health, which already in 1932 was the reason for his voluntarily giving up his professorial chair, certainly worsened through all that took place at the University of Marburg from 1933 on. The dismissal of Jewish and half-Jewish professors, including the outstanding physician Professor Ernst Freudenberg, and also several "Aryan" theologians whose political views were too far to the left, was certainly a hard blow. Hermann Jacobsohn, professor of Indo-Germanic linguistics, who was a dedicated advocate of a German-Jewish symbiosis in the sense of the Central Association of German Citizens of the Jewish Faith (*Centralverein Deutscher Staatsbürger Jüdischen Glaubens*), committed suicide by throwing himself in front of a train in April 1933. So too, the death of Rudolf Otto, who felt he shared the responsibility of his people, was overshadowed by the political events of that sinister time. After an accident in his museum he died on March 6, 1937. I wrote an obituary in the Hebrew daily *Ha'aretz*, paying tribute to the eminent researcher and long-time friend.

As often in life, very trivial events sometimes have serious consequences. That happened to me in the spring of 1923, when together with my wife and oldest daughter I was visiting my in-laws in Amsterdam. One day I received news from Dr. Leo Polak—then *Privatdozent* at the City University of Amsterdam and later professor of philosophy at the (Reich) University in Groningen before becoming a victim of the Nazis—that I should immediately contact Heinrich Scholz, a professor of philosophy in Kiel who was temporarily in Amsterdam. I had visited Polak a short time earlier and left him, as he requested, some proof sheets from the second edition of my *Einleitung in die Mengenlehre*, which then appeared half a year later in Berlin. The proofs in question, about which Polak evidently told the Kiel professor, were about Brouwer's intuitionism; the logicism of Bertrand Russell, one of the greatest logicians since Aristotle, perhaps with the exception of Gottfried Wilhelm Leibniz; and the formalism of Hilbert. Now Scholz definitely wanted to get to know the unfamiliar author—that is, me—and get the chance to talk. We did in fact meet a short time later, and the meeting developed into a scientific and personal friendship, which continued without interruption until Scholz's death. His last letter to me in Jerusalem before the war started was dated August 1939, and despite his serious illness in the years following the war, his publications and letters started coming again from 1949 on. Our last personal meeting goes back to the period before the Second World War, when he could still take part in the Ninth International Philosophy Congress in Paris in 1937 (Congrès Descartes) and I had the honor of holding one of the keynote lectures on "Discontinu et Continu." At the Tenth Congress in Amsterdam in 1948, to which I had traveled from the recently liberated Jerusalem, I met with Bertrand Russell; I looked for Scholz to no avail. His declining health had not permitted him to travel.

Heinrich Scholz

Heinrich Scholz was born in 1884 in Berlin. His father was the Brandenburg "Privy Consistorial Councilor" Hermann Scholz, a student and friend of the leading theologian Adolf von Harnack, who by the way was also greatly influenced by Rudolf Otto. A Protestant theologian, Heinrich Scholz became professor of the philosophy of religion in the Protestant theology school at the University of Kiel in 1917, when he was just 33 years old. In 1921—the year his major theological work *Religionsphilosophie* was first published—he traded this position for a philosophy chair at the School of Arts and Humanities (*Philosophische Fakultät*) at the same university. In 1928 he changed again, assuming a professorship at the University of Münster, after the ministry had satisfied all his requirements for basic research. At first he had been offered a chair in philosophy, but later, during the Nazi period, it was changed into one for mathematical logic and basic research and assigned to the School of Mathematics and Natural Sciences. This made it possible for him to avoid a politicization of the instruction. He also invited foreign scholars to Münster, such as the outstanding Polish logician Jan Łukasiewicz, who at the time was still a professor in Warsaw, but who started teaching in Dublin in 1949.

This is not the place to explain the unique evolution of an academic teacher at a mature age from theology to philosophy to mathematical logic, actually mathematics. Suffice it to quote a few sentences from a lengthy letter of his lifelong friend, the philosopher Eduard Spranger, from December 1957: "The causes for the changes were to be found at the depths of his soul. Scholz lost his gracious first wife in 1924. At his wife's grave . . . he said to me, 'You understand how I am no longer able to work on things of substance.'"[38]

[38]See *Heinrich Scholz*, Schriften der Gesellschaft zur Förderung der Westfälischen Wilhelms-Universität zu Münster, no. 41 (1958).

But his change to "formal" things certainly did not mean settling for less substance, as revealed by the enthusiasm that inspires his logical, mathematical work. To explain this let me cite just one sentence from his 1934 treatise "Warum Mathematik?" (Why Mathematics?), in which he first rejects the thesis of the broad applicability of mathematics and that of the ennoblement of humanity (Fichte) and then continues: no, mathematics is great "for the sake of those who are so enlightened by its beauty that the question ceases to exist for them." The "form" or "structure" was the pinnacle for him—also in the landscape, when in the highlands the trees are left behind and the bare, hard rock becomes visible. This is the only way to understand how in the 1920s in Kiel, despite his failing health, which persisted until his death, he again took a seat in the classroom among the young students and completed a full course of study in mathematics and mathematical physics. Then he continued to work together with mathematicians and physicists in Kiel and Münster, also with a generation of younger researchers whom he himself had trained and with whom he elevated the Münster School to international esteem.

In June 1925, I accepted his invitation to give a course on the foundations of set theory for the Kant Society in Kiel. The course comprised ten lectures which appeared in book form in 1927. My scientific correspondence with Scholz that had been ongoing since 1923 intensified after this resumed personal contact. When in fall 1927, Otto Toeplitz, who was a mathematics professor in Kiel together with Ernst Steinitz, accepted a position in Bonn for the following spring, Scholz spoke out in the appointment commission selected by the School of Arts and Humanities, requesting that my name be included on the list of candidates for the vacant chair, which would be submitted to the ministry. Even if Steinitz and Toeplitz supported him in this endeavor, it was presumably Scholz's energy and influence that was decisive. The School of Humanities did in fact accept the list of recommendations in which my name appeared in second position, but the ministry felt that the professor heading the list, Helmut Hasse, was indispensible to the University of Halle at that time.

I received notice of my appointment, which was almost thwarted by Brouwer's negative opinion of me, in March 1928, and began my teaching duties as a full professor in Kiel in late April with a lecture on "the essence of mathematics." Scholz was among the most attentive in the audience. He had received his position in Münster starting in February 1928, but he arranged to be able to remain in Kiel for the summer semester. The value of this extraordinary personality and the significance of our relationship, then and later, for my further career, come back to me clearly as I re-read our letters.

Finally I would like to make some remarks about my *Jewish* activities in the years from 1919 to 1929 as well as the beginning of my work for the Hebrew University. Professor Moritz Sobernheim, who headed the Jewish affairs section in the Foreign Office of the Weimar Republic, convinced me to write a memorandum in 1919 "about the demands pertaining to legislation in the national home of the Jewish people from the standpoint of Orthodox Jewry," which was to be presented to the Peace Conference in Versailles. My initial refusal, since I felt incompetent for the task, was countered by his insistence. I do not know what has become of the

19-folio-page memorandum—except for the fact that it was presented to several rabbinical and Zionist offices. It presumably found its justified end in the waste-paper basket. Incidentally, it was also due to Sobernheim's efforts that the Foreign Office prompted the Prussian Ministry of Culture to grant me a leave of absence for two years so I could go to Jerusalem.

In Marburg, where Dr. Naphtali Cohn[39] served as rabbi after the war, Jewish life was not very palpable, despite the Orthodox leadership in the institutions of the Jewish community. Undemocratic charters and a non-spiritual atmosphere made it difficult for us to get settled there. However, we received suggestions from the religious Jewish students, of which there were about a dozen studying in Marburg at any one time. Many of them came from the rabbinical seminary in Berlin in order to complete their doctorate in Marburg or nearby Giessen, where Dr. Israel Rabin and later Dr. Samuel Bialoblocki were lecturers in post-biblical Judaism and modern Hebrew, and then to become rabbis in various Jewish communities throughout Germany. On Friday evenings there were almost always individual students as guests in our home; most of them came again on Saturday afternoons. Sometimes I would give a class on the Gemara or a lecture on one topic or another. I developed particularly close relations with the excellent Lithuanian Talmud scholar Moses Silberg, who had studied law in Marburg and as of 1950 was an Israeli Supreme Court justice. For a long time he was my teacher in Talmudic literature.

Dr. Aron Barth

Soon after the war, I was chosen along with my friend Dr. Aron Barth to serve on the board of the Berlin rabbinical seminary, which was presided over with great skill by Hermann Gumpertz of Hamburg. My father had been a member of the board for decades. The annual meetings were generally very interesting. It proved impossible to find an equally worthy successor to Professor Jakob Barth, but the

[39]His wife Hanna was a daughter of Selig Auerbach.

main Talmudic chair, which had been vacant since the death of Professor David Hoffmann in 1921, enjoyed an unparalleled, although astonishingly short, heyday: The board decided to introduce the Lithuanian style of study and teaching, which was hardly known in Germany. Consequently, after consultation with Rabbi Yosef Shlomo Kahaneman, the "Ponevezher Rav," who was a recognized authority, the young Rav Avraham Eliyahu (Elya) Kaplan was given a professorial position as a Talmudist at the rabbinical seminary.

Rav Avraham Eliyahu (Elya) Kaplan

Kaplan was an absolutely unique personality: a first-rate Talmud scholar and authority, specifically matured in the Slobodka yeshiva, who knew how to harmoniously unite Lithuanian acumen with Western scientific methods. He was also a top-notch author and journalist, who never let his independent judgment be dimmed by any Jewish-political influences. On top of that he was a poet who wrote in Hebrew, which he mastered amazingly well—his native tongue was Yiddish—as well as a cultural politician and much more.[40] When at board meetings he brought up plans of the undertaking, which was supposed to have become his life's work, and which related to a commentary on the Babylonian Talmud that was appropriate to our time, you would think you were listening to a prophet. We board members and especially his profoundly impressed students regrettably did not have the good fortune to enjoy him for very long. He died suddenly in the middle of his work in 1924, when he was only 35 years old, the same age at which his father had died, before his son was even born. In my entire life I never again met a Jewish personality of his caliber, except for Rav (Abraham Isaac) Kook.

[40]A selection of his works, insofar as they were not of a specialized, Talmudic nature, appeared under the title בעקבות היראה [*B'ikvot HaYirah, In the Footsteps of Yirah* (Awe)] (Jerusalem: Mossad Harav Kook, 1960).

Of the other Jewish scholars with whom I had contact in those years, I would like to mention briefly three who made a particularly deep impression on me, including Adolf Büchler, director of the Jews' College in London, and Immanuel Löw, Orientalist and rabbi of Szeged, Hungary (who had been imprisoned several times in 1920–1921, during the White Terror). I admired the two of them for their Jewish scholarship and the way in which they were able to convey their knowledge and original ideas in such a fascinating fashion. I met with both of them frequently in Switzerland; with Büchler also in England. They were the only ones whose company took precedent for me over a mountain hike that I had already planned. I remained close to Büchler until he died, also through our cooperation on the Board of Governors of the Academic Council of the Hebrew University, of which he was among the most conscientious and energetic members for many years.

My visits with Franz Rosenzweig in Frankfurt am Main were experiences of a very different kind. After he completed his dissertation on "Hegel and the State," which was published in 1920 as a greatly expanded two-volume work *Hegel und der Staat* (rev. ed. 1962), he went through difficult internal struggles and finally decided against completing a postdoctoral professorial qualification (*Habilitation*) and instead to put all his energies into the Jewish cause. While still a soldier, in 1917 he addressed an open letter to Hermann Cohen titled "Zeit ists" [It's Time!], which contained his "thoughts on the current problems of Jewish education." In 1920, shortly before his magnum opus *Der Stern der Erlösung* (*The Star of Redemption*) was published, he founded the Free Jewish School (*Lehrhaus*) in Frankfurt, which he directed the first two years until he became ill. I admit that it was largely his personality that attracted me to him, because at the time of my visits, his affliction[41] had already worsened to such a degree that he could speak to me only through a device operated by his wife. He nevertheless remained tirelessly involved with his writing and scholarship. The teaching position for which he was originally considered, to hold lectures on Jewish religious studies and Jewish ethics at the University of Frankfurt, could not be given to him after all, and Martin Buber received it instead in December 1923. However, the two friends jointly began a German translation of the Bible in 1925, which Buber finally completed decades later in Jerusalem. It was a heroic struggle that for many years was successfully and productively guided by the spirit of an utterly extraordinary person with a serious illness!

I made statements on Jewish ideological questions in lectures and occasional articles for journals. The only essay I would like to mention here is the one I dedicated to my deceased father, "Leben, Natur, Religion" [Life, Nature, Religion], which appeared in the October issue of volume 12 (1925) of the Berlin journal *Jeschurun*. This essay was based on the book with the same title by Friedrich Dessauer, a sophisticated biologist and physician, who as a "non-Aryan" lost his

[41]Amyotrophic lateral sclerosis (ALS).**

professorship at the University of Frankfurt, also because he belonged to the left wing of the Catholic Center Party. He found a professorial chair in Switzerland.

At this point I shall briefly depict my very superficial experiences with anti-Semitism in the 1920s. With respect to the universities, the anti-Jewish tendencies at that time—as opposed to the situation prior to the revolution—were far more pronounced in Bavaria than in northern and western Germany. The appointment, though not the promotion, of Jews to positions as full professors remained rare with the exception of the new city universities in Frankfurt and Hamburg. Regarding Munich in particular, the "capital of the movement," much has been written about the sensational academic scandal, which started when the Jewish-Norwegian crystallographer Victor M. Goldschmidt did not receive a professorship in Munich and ended in 1925 when the chemistry Nobel laureate Richard Willstätter, who was a friend of Chaim Weizmann, resigned in protest. Nevertheless, the outstanding physical chemist Kasimir Fajans received a promotion to professor around the same time in Munich.

I could not complain about personal anti-Semitic experiences either in Marburg or after I received my professorship in Kiel—although in the last quarter of the nineteenth century Marburg and surroundings were a stronghold of anti-Semitism. The dean of the School of Arts and Humanities (*Philosophisches Fakultät*) in Kiel even told me expressly in January 1929 that the School "has decided to make every effort to free you from all department business every Saturday as well as on all your holidays."

A poisonous anti-Semitism surfaced very early in Germany's and Austria's Alpine Associations, in which membership was practically required for mountain climbers in Bavaria and the Austrian Alpine regions. As early as 1924 I had a very unpleasant correspondence with the Munich section of the German Alpine Association, which resulted in my cancelling my membership a short time later. Pro forma I became a member of the Marburg and later the Kiel sections. Not until 1933, when I was in Jerusalem, did I get word from the Kiel section, saying that I was being removed from the membership roster, which came with an expression of regret at having to lose such a loyal member after 24 years in the German Alpine Association as a whole. At the same time I was informed that the decision was final, since it was based on the "Aryan Paragraph," the new law that precluded Jews from club memberships.

Regarding how the poison of anti-Semitism affected a Jewish fate, I can list many examples from the academic world that I more or less personally witnessed. I will describe one case that is not entirely typical in terms of time and place, but is nonetheless worth mentioning. Since the turn of the twentieth century, Tullio Levi-Civita has been considered one of the greatest mathematicians in Italy—among whom the Jews comprised a greater share, both quantitatively and qualitatively, than anywhere else, perhaps with the exception of Hungary. As of 1902 Levi-Civita was a full professor, teaching in Padua, and in 1918 he went to Rome to assume a comparable position. His research on the "methods of absolute differential calculus and its application"[42] played a crucial role in the expansion of the general theory of

[42]*Mathematische Annalen* (Springer) 54 (1–2) (March 1900): 125–201.**

relativity. He was a member of many academies and had an honorary doctorate at numerous universities. I knew him personally for a long time and met him many times at conferences as well as on the Academic Council of the Hebrew University, of which he was a member, and later also at his home in Rome. Like his other Jewish colleagues he suffered twofold under the fascist regime. In 1931 he, as all professors, was supposed to swear an oath to the fascists, which he could not reconcile with his conscience. That alone led to his losing many of his positions and to restrictions in his research activities. In 1938, due to the race laws that Mussolini imposed on pressure from Hitler, he was finally removed, along with all Jews, from Rome's university and academy, but not from the Pontifical Academy of Sciences. Both times I offered him new opportunities, in 1938 even at the Hebrew University, with the approval of Chaim Weizmann. And both times he was initially unable to decide and, after long deliberations, declined the offer. He died in 1941, at the beginning of the war, at only 68 years of age. His letters that I have saved, which also contain copies of his related correspondence with the rector of the university in Rome, are moving testimonials of emotional greatness and strength of character. That same year, 1938, incidentally, the brilliant physicist Enrico Fermi, whose wife Laura was Jewish, did not return to his professorship in Italy from Stockholm, where he went to receive the Nobel Prize. Instead he emigrated to the United States and became a leading figure in nuclear research.

I participated in only a few Zionist meetings. In December 1918 I spoke in Berlin at the 15th Delegates' Convention of the Zionist Federation of Germany (ZVfD), which focused on the talks given by Kurt Blumenfeld and Salman Schocken. In January 1920 I was the secretary of the World Conference of the Mizrachi in Amsterdam, where, aside from the painter and etcher Hermann Struck, my friend Aron A. Barth, and the Orientalist Hermann Pick, I also met, for the first time, Rav Meir Berlin, the Sejm representative Joshua Heschel Farbstein, and Rav Yehuda (Judah) Leib Fishman (Maimon), who from 1948 to 1951 was Israel's Minister of Religions. I presented a resolution relating to the Hebrew University, which despite the initial opposition was ultimately accepted almost unanimously. In September 1921, due to a Polish Mizrachi mandate, I participated as a delegate in the first postwar Zionist Congress in Karlsbad (Karlovy Vary), in which the Arab question was broached, namely, by Martin Buber.[43] I must admit, I found the convention proceedings and in general all the politicking, but especially the demeanor of the Mizrachi leaders from the United States and Poland, so unpleasant that after that I avoided Zionist conferences as best I could. Still, at a special conference of the Zionist Federation of Germany in 1931, I gave the keynote talk on "Hebrew education and culture in Palestine."

[43]See Martin Buber, *Der Jude und sein Judentum,* collected essays and speeches, with an introduction by Robert Weltsch (Cologne: Joseph Melzer, 1963), 474–487 [English: Martin Buber, "A Proposed Resolution on the Arab Question" (Sept 1921) and "Resolution on the Arab Question of the Twelfth Zionist Congress" (Sept 1921), in *A Land of Two Peoples: Martin Buber on Jews and Arabs*, chapters 5 and 6, ed. and trans. Paul Mendes-Flohr, (Chicago: University of Chicago Press, 1983), 58–63**].

Some good things came of my participation at the conference in 1921. Among the many Mizrachi delegates from Eastern Europe, I especially noticed Rabbi Jakob Hoffmann from Rădăuţi (Romania) because of the logic of his argumentation and his sense of order, which fit together well, as I inferred, with eminent Talmud scholarship. In 1922, after the death of the Orthodox rabbi of the Frankfurt congregation Dr. Nehemia Anton Nobel, I received an official request from Frankfurt asking which of the three proposed candidates for Nobel's successor I would recommend. I decided on Hoffmann. He was in fact selected and he fulfilled the duties of his office with great success until he was arrested by the Nazis in 1937. He managed to escape to New York a year later, where he continued to serve as a rabbi for many years.

The German Mizrachi Organization, under the leadership of Aron Barth and Hermann Struck, who moved from Berlin to Haifa as early as 1922, became increasingly independent of the World Mizrachi Organization starting around 1928. At first I joined this "independent" Mizrachi group, but soon after resettling in the Land of Israel/Palestine, I transferred to the HaPoel HaMizrachi (lit. "Mizrachi Workers"). After the founding of the State of Israel, I broke off with them as well as I did not think "religious" parties in an independent state were in keeping with the times. I felt much better being non-affiliated and independent.

In addition to being politically active, in the 1920s I was also more involved in Zionist journalistic activities. I started by writing essays on "questions of sexuality in the formation of Palestine" (*Jüdische Rundschau* and *Jüdische Presse* 1919), wherein I dealt with questions of demographic policy, including tax benefits for large families. Later I wrote more about cultural and educational issues following Zionist or Mizrachi conferences. Aside from the *Jüdische Rundschau*, which was edited in an absolutely exemplary way by Dr. Robert Weltsch, there were also some weeklies that were closely affiliated with the Mizrachi, such as the *Jüdische Presse*, and from 1924 on, the *Jüdische Wochenblatt*. The polemics of the anti-Zionist Orthodoxy,[44] which were at home in the Agudah and in Jacob Rosenheim's *Israelit*, assumed a considerable role in these publications. A typical example of the truly incredible excesses that these polemics could lead to was the proposal made by a northern German rabbi in 1924 in the *Israelit*, which the editors printed without any commentary. This really superb suggestion was that since the Zionists had monopolized Palestine/Land of Israel as the Jewish homeland, the Orthodoxy should found a corresponding homeland in Babylonia, the land of provenance of the Talmud.

[44]An article promoting cooperation between the Agudah and the World Zionist Organization, which was written by the loyal Agudist Dr. Moses Auerbach, who spent his life working for the Land of Israel, was printed in the back pages of the *Israelit* of February 12, 1921, as previously the news of the resolution in San Remo (where it was resolved to include the Balfour Declaration in the peace treaty with Turkey and to transfer the mandate for Palestine to Great Britain) appeared under "Club News." On the other hand, the lead article in the same issue was dedicated to a rendering of an anti-Zionist essay from the *Jewish Chronicle*, which rejected the British draft mandate for Palestine.

Very quickly, though, my writing activities started focusing more and more on the Hebrew University. It is worth going into greater detail, especially in view of the change in the spiritual stance toward the university among the Orthodoxy from Germany.

In fall 1919 I received a mailing in Hebrew from the London central office of the World Zionist Organization signed by Shmaryahu Levin, inviting me to a conference of Jewish scholars in Basel in January 1920. The conference was organized to plan the structure of the Hebrew University in Jerusalem. The letter made reference to the resolution of the Eleventh Zionist Congress in 1913 in Vienna, which called for the institution of a founding commission, and to the laying of the foundation stone by Chaim Weizmann that took place on July 24, 1918, in Jerusalem.[45] This announcement mentioned schools of medicine, law, natural sciences, sociology-economics, and Jewish and philosophical-philological subjects. In view of other obligations I was initially unable to confirm my participation in Basel, and in the end the conference never took place.

In connection with that, I corresponded from January to March 1920 with Professor David Hoffmann, rector of the Berlin rabbinical seminary. Recognized by most of the Orthodox community as a leading scholar, he told me his opinion on how to avoid a "culture clash" at the Hebrew University. Because the letters he wrote me shortly before his death, which demonstrate how profoundly he considered the problems, were later publicized, there is no need to go into detail about them here.

In the years 1924 and 1925, before and directly after the official opening of the Hebrew University on April 1, 1925, I expressed my views on the problems facing the university, its structure, its spiritual direction, and the treatment of Jewish subjects, not only in articles in German and English, but also for Orthodox institutions that asked for my opinion. I received an extensive telegram, for example, from the Amsterdam Jewish Community, which then in response to my answer sent congratulations to Jerusalem, causing some inconvenience for several congregation leaders in Amsterdam. The struggle within the Orthodox press in Germany reached its peak in early 1925, when a long, unsigned editorial in the Frankfurt *Israelit*—as it later turned out, written by Rosenheim—drew the conclusion that it was "banned according to religious law for every conscientious Yehudi [Jew] to directly or indirectly promote the university." I expressed my amazement in the *Jüdisches Wochenblatt* of February 6 and 20 that a newspaper editorial office felt called upon to make that kind of decision on religious law. This prompted, among other things, the letter of a leading and highly educated member of the Frankfurt secession congregation, that spoke of "your infamous article," "your shameful attack on the

[45]Here I would like to comment that the location of the university (from 1925 to 1948) is generally referred to as הצופים הר (Mount Scopus), which is actually the northern extension of the ridge that is the Mount of Olives mentioned in the Bible. In geographic and literary terms there is not even a Mount Scopus, but only the Scopus (צופים) mentioned in Talmudic literature, which describes the ancient watershed road from Jerusalem to Nablus (Shechem) and lying much farther westward than the university building.

honor of Mr. Rosenheim," and the author of the letter "sincerely regretted ever having received you in my home," after having invited me a short time earlier to a Sabbath meal. In that he was more pope-like than "Pope" Rosenheim, with whom I met on friendly terms that same year. In any case, this and similar incidents show the rancor with which the ultra-Orthodoxy opposed the Hebrew University up to Hitler's coming to power. It even almost caused a crisis within the board of the rabbinical seminary in Berlin, because its most influential member characterized my article about the Hebrew University in a private letter—which he had not intended for me to make public—as the "height of baseness," and "evidence of heinous sentiments." And when I accepted the professorship in Jerusalem in 1928, even such a meritorious and otherwise calm Agudah leader as Dr. Leo Deutschländer in the Vienna group of the Association of Jewish Academics, which on occasion had wanted to invite me to give a lecture in a class at the University of Vienna, said that by accepting the position in Jerusalem I had excluded myself from Orthodox Judaism.

Already in spring 1926, I had taken a four-week trip to the Land of Israel/Palestine with my wife and my brother, and we saw virtually every corner of the country, including the Arab regions. We spent Passover in Jerusalem. I got to know numerous leading personalities of the *yishuv*, the body of Jewish residents in Palestine/Land of Israel, in particular the very diverse rabbinical leaders Abraham Isaac Kook, Jacob Meir, and Chaim Sonnenfeld,[46] and spent many hours with German emigrants in the *yishuv*, especially with my friends Dr. Moses Auerbach, Professor Hermann Pick—then member of the Palestine/Land of Israel Zionist Executive—and Hermann Struck. At the interesting home of David Yellin in Palestine/Land of Israel, I also met the sociologist and economist Professor Franz Oppenheimer of Frankfurt, who was there on a study tour and who voiced harsh criticism of the new developments in the kibbutz movement. Although at the time there was no mention of any teaching position at the nascent university, I was invited to give a guest lecture on a mathematical topic there. I presented the introduction in Hebrew, but the main part of the talk in German, since I lacked familiarity with the—then still nonexistent—Hebrew terminology in higher mathematics. I also received an offer at that time: a Mizrachi teachers' seminary with an excellent teaching staff offered me a position as a mathematics teacher. After some deliberation, however, I could not make the decision to give up my research and the training of scientists.

Back in Kiel I unexpectedly received the offer of a professorial position in Jerusalem. I corresponded with Chief Rabbi Kook before I made my final decision. He was perhaps the last rabbinical leader of great caliber and independent attitudes that Judaism has ever known. He responded quickly with a telegram and more extensively in a letter, in which he explained that I should without a doubt accept

[46]The novel by Arnold Zweig provides a good introduction into the world of Jerusalem's extreme Orthodoxy: *De Vriendt kehrt heim* (Berlin: Gustav Kiepenheuer, 1932) [English: *De Vriendt Goes Home*, trans. Eric Sutton (New York: Viking, 1933)**].

the position, despite all the reservations that he himself had expressed against the university. The presence of religious men in the teaching faculty was the best guarantee for desirable developments, he said, remarking that his advice was also sent on behalf of my friend Rav Simcha Assaf, who later too was a professor, of Talmud studies, at the Hebrew University.

I desperately needed such moral support. In June my sister-in-law Meta Posen wrote how "terribly" worked up people in the Breuer family circles and beyond, in broad circles of the "religious community," were getting over the idea that I might accept the position in Jerusalem. The same was true for the Berlin Orthodox community. No one could guess at that time that only a few years later the Orthodox sons, grandsons, and nephews of these fervent protesters would become students and teachers at the criticized university. One could guess even less that only ten short years later some of those attackers were to send telegrams to me, then rector of the Hebrew University, beseeching me to send them student certificates that would enable them to leave Germany and remain alive!

The fact that we had already left Germany in 1929 cannot be attributed to any special foresight, but first, to the Jewish and Hebrew education that I had enjoyed since my early childhood, second, to the Zionist attitudes of my wife and myself, and third, to the lucky circumstance of my receiving a professorship in Jerusalem in 1928. For us this did not mean the absurd emigration of a German university professor "into the desert," as people around me saw it, but rather a dream come true. Regarding my relocation as a professor at the Hebrew University, as a motto I could use a sentence taken from a professor of medicine whom I know: "A life is meaningful and fulfilled when it has the options to make use of the multifaceted potential it possesses and can be freely shaped according to these options."

Chapter 6
Epilogue (1929–1933)

I would like to mention here only a few facts about my activities at the Hebrew University during my teaching years 1929–1931. My students were indeed very talented but rather few in number. Still, they made up 10–20 percent of the entire student body of around 200 at that time, and their interest gave me great satisfaction. On the other hand it did not take long before I too recognized the weaknesses of the administration that had pushed Einstein and later Weizmann into the opposition.

The landscape and our life in a new environment captivated us profoundly. Learning to use the Hebrew language even in everyday usage came easily to me and the children. On extended trips that I took with my students and friends throughout the country, all the way to the peak of the over 9000-ft-high Mount Hermon and the fairytale-like desert city of Petra, we learned about the life of the Arabs, the Druze, and even the Bedouin, at least superficially.

There were relatively few German Jews in the country and especially in Jerusalem. Some taught at the university. We had all the more contact with Russian and Polish, and some American, Dutch, and Baghdadi Jews; at that time German—more so than English—served as a means of communication in addition to Hebrew. During our summer trip to Europe in 1930, Einstein prompted me to give him a comprehensive report on the university. On this vacation I also participated in the Seventh World Congress of Philosophy in Oxford and got to meet the philosophers and mathematicians who had come from the United States: Morris R. Cohen, Paul Weiss, and Alfred North Whitehead; I was particularly impressed by Cohen. Whitehead, a mathematician who was also one of the most significant philosophers of his time, was already almost 70.

My only disappointment in my field in Jerusalem was the fact that my request for an applied mathematics chair, approved by the University Council in early 1930, was brusquely rejected by the university management, although the two potential candidates I proposed, Dr. Sydney Goldstein and Dr. Louis Rosenhead, both from England, had already made brilliant names for themselves, despite their youth. Both later became Fellows of the Royal Society—the highest scholarly distinction that a British mathematician can receive, as there is no Nobel Prize for mathematics. The

A.A. Fraenkel, *Recollections of a Jewish Mathematician in Germany*,
DOI 10.1007/978-3-319-30847-0_6

establishment of a chair for applied mathematics was not rejected for financial reasons, as the effects of the New York stock market crash in late 1929 were not felt by the university until a year later. Instead, the gentlemen in the management felt that two mathematicians in senior positions, plus a lecturer, would represent a hypertrophy of mathematics.

In spring 1931, in view of the fact that my leave of absence from the University of Kiel was coming to an end, the time had come to formulate my conditions for extending my stay in Jerusalem. Since our children had acclimated themselves entirely, my wife and I agreed to stay if we could. I had no intention of asking for a salary increase. My "conditions" were soon publicized in the context of the investigative commission of 1933–1934; they were barely significant except for the basic prerequisite that two senior positions in mathematics be retained. I could not take the risk that my remaining would trigger the dismissal of my outstanding colleague Michael Fekete, who had no option of returning to the University of Budapest on account of political circumstances. In addition to this personal reason there was also a practical one: Mathematics with only one senior position did not make much sense in Jerusalem. Gifted students could only be furthered if they had good instruction. In fact, however, my chair was not filled again after I left. The financial situation of the university in 1931 did not permit it. For this reason in particular the leadership in Jerusalem decided to reject my conditions, to end my contract in the fall of 1931, and even to reject my request for a leave of absence. Luckily, the Board of Directors, which convened in summer 1931 in Zurich, showed greater understanding. It allowed me to remain on the faculty of the University in Jerusalem and gave me the status of professor on leave, without pay.

Consequently, we returned to Kiel in late October 1931 for another three semesters. I have few memories of this time. The academic atmosphere was dampened, not only because of the politicization of the students. Particularly painful for me were the changes in the department of philosophy, which was very close to my heart. The sophisticated Julius Stenzel, who together with Toeplitz had been interested in Greek mathematics, still taught there, but he conveyed a rather aged impression and his position was endangered on account of his having a Jewish wife. We also strongly sensed how much the second chair in philosophy—which up to 1928 had belonged to Heinrich Scholz—had become alienated from its tradition. There was a minority opinion in the humanities school (*Philosophisches Fakultät*) in 1929, which had recommended Rudolf Carnap—who later became a world-renowned logician—as first on the list. But the ministry kept to the majority opinion and gave the position to the Hegelian Richard Kroner. Kroner—nephew of the Württemberg chief rabbi (carrying the title "Ecclesiastical Council Member") Theodor Kroner, whom I knew well—was no longer Jewish and even attempted to conceal his background from his colleagues and students, while representing a pronounced "German" philosophy. That was of no help to him in 1933, when in the fall of that year nationalist students demanded his dismissal.

Through school, our children became friends with the family of the German literature Professor Carl Wesle. When the Nazis took power in 1933 it turned out that Wesle had long been a member of the Nazi party. As such he was promptly made dean of the School of Arts and Humanities. When I resigned from my professorship in Kiel in April 1933, he sent me an official letter of thanks and of farewell in his position as dean, but also attached a very cordial, personal farewell letter. Unbeknownst to me, and of course to no avail, he had turned to the ministry in Berlin and applied for a pension for me. (According to official records, I had accumulated 19 years of service, including my years of war service). When I acknowledged having received his letters I expressed my amazement that a leading Nazi would go to such efforts for a Jew. He responded that the Party had nothing against me, as I had always openly referred to myself as a Jew.

My last year in Germany was not particularly calm given the political circumstances, but it was not very significant either. When invited to scientific lectures in March 1932 in Vienna, I took advantage of the opportunity to talk in Vienna's "Urania," a public educational institute, about the Jewish efforts to build up Palestine/Land of Israel, and toward the end of the year also in Breslau's Zionist Organization, when my lectures naturally always focused on matters of education.

I consider my correspondence in summer 1932 with Chaim Weizmann to be significant; handwritten at times, it is now preserved in the Weizmann Archives. In addition to special research matters, our communication revolved around the Hebrew University. He urgently requested—and I reluctantly agreed—that I participate in a session of the Board and the Academic Council of the University that was planned for August in London. Its purpose was to support Weizmann himself, along with Selig Brodetsky, Leonard Ornstein, and others, in pushing through reforms in the leadership in Jerusalem. Nothing came of the session due to resistance from the United States and Jerusalem. Weizmann also asked me to return to Jerusalem—which corresponded with my wishes and those of my wife, and was also the better fit with regards to the terms of my leave of absence.

In early 1933 developments followed in rapid succession. Although I continued teaching until the end of the winter semester, I left Germany forever as soon as the semester ended in late February. On the day of my departure, at the Frankfurt train station, I was able to give my future friend Dr. Yitzhak (Ernst) Nebenzahl some tips about emigrating to Palestine/Land of Israel. This highly gifted man's decision to emigrate would have a brilliant outcome for him as well as for the Jewish community of the *yishuv* and the State of Israel. A few years ago, in 1961, he was elected by parliament to be Israel's State Comptroller.

My trip took me first to Switzerland, where I had long been planning to give lectures. Thanks to the far-sighted invitation of my brother-in-law Hugo Prins, my wife and children had already left in March to stay with my mother-in-law in Amsterdam, where we all enjoyed several months of hospitality. From there I sent my letter of resignation to the University of Kiel in April, requesting release from my contract.

I spent summer 1933 in Holland and London, partly engaged in scientific work and lectures, and partly in the aforementioned negotiations with Einstein, together with Weizmann, Abraham Yahuda, and others, which ultimately led to Einstein's reconciliation with the Hebrew University. I received a number of invitations, more or less paraphrases of each other, namely from Scandinavia and from a university in the French part of Switzerland, but I turned them all down in view of my external and internal ties to the university in Jerusalem. Only at the University of Utrecht, in Holland, did I give lecture classes over the course of two months, at the invitation of the eminent mathematician Julius Wolff, a Jew who shortly before the end of the war became a victim of the Nazis. Among my students there was the mathematician and logician Evert Willem Beth, whose progress was exceptional and who later became a professor at the University of Amsterdam.

Also in Utrecht that same summer, a rump board meeting of the Hebrew University was convened with the physicist Leonard Ornstein presiding and including chancellor Dr. Magnes. With Magnes's agreement it was determined that I should resume my teaching activities in Jerusalem in the 1933–1934 academic year. That fall we returned to Jerusalem, to live there permanently and happily.

Chapter 7
Afterword: 1933–1965
by Jiska Cohen-Mansfield

Originally, Abraham (Adolf) HaLevi Fraenkel had planned to write a three-volume autobiography,[1] of which this would have been the first volume. He died prior to writing the subsequent two volumes, the subjects of which were to be his life after leaving Germany and immigrating to Israel in 1933. The first volume, presented here, was written in German, whereas he intended to write the latter two in Hebrew. The second volume was intended to cover the years from his arrival in the Land of Israel (Palestine) in 1933 until the establishment of the State of Israel in 1948; the last volume would have covered the years thereafter. This afterword summarizes the events of his life after 1933.

Academia—The Hebrew University

Fraenkel's initial involvement with the Hebrew University is detailed in Chap. 5. He came to teach at the Hebrew University in 1929 and left at the end of the 1931 winter semester to return to Kiel University, which had granted him a two year leave of absence. He intended to return to Jerusalem in the future. Indeed, many letters to Fraenkel from professors and lecturers at the Hebrew University between 1931 and 1933 reflect his continued involvement with Hebrew University affairs during his stay in Kiel. These affairs included the development and evaluation of existing departments, recruitment of new faculty, discussion of the financial affairs of the *yishuv* (the Jewish settlement in the Land of Israel/Palestine) and means to assist it, and various other matters. In 1933, Fraenkel returned to the Land of Israel where he remained for the rest of his life.

Fraenkel's appointment was the first Hebrew University permanent appointment of a leading scholar with a worldwide reputation. He was the only full professor who already had that rank in a university abroad, at the University of

[1]Moshe D. Herr, personal communication.

A.A. Fraenkel, *Recollections of a Jewish Mathematician in Germany*,
DOI 10.1007/978-3-319-30847-0_7

Kiel.[2] Fraenkel was involved in all aspects of the Hebrew University development. In aiming to elevate the university to international stature, he sought out prominent academics and advocated appointing those fleeing Nazi persecution. He was involved in developing university policy, such as requiring Hebrew proficiency as a condition for employment and promotion (enshrined in the Senate protocol of 1st January, 1941). In addition, he was involved in developing the university campus.

Fraenkel's leadership of the Institute of Mathematics at the Hebrew University elevated it to one of the most important of its kind worldwide. Once Fraenkel arrived, the department of mathematics was the only department with a complete curriculum and a full roster of lecturers. As the Institute of Mathematics initially belonged to the Faculty of Humanities (the only faculty during those early years), Fraenkel became its first dean (1929–1931). When a School of Sciences was established and the Institute of Mathematics eventually incorporated into it (hence: the Faculty of Mathematics and Natural Sciences), he became its first acting dean (1935–1938). (Chaim Weizmann, who had no university appointment at the time and who later became the first president of the State of Israel, was formally nominated the first dean of that school, but he assigned the position to Fraenkel).

Magnes and Fraenkel

[2] כהן, אורי מוסדות ניהול האוניברסיטה העברית 1925-1948: תהפוכותיה של החתירה לאוטונומיה מוסדית מוחלטת. מתוך: תולדות האוניברסיטה העברית בירושלים התבססות וצמיחה א' עורכת חגית לבסקי. הוצאת מאגנס ירושלים 2005. Cohen, U. Governing Institutions of the Hebrew University 1925-1948, In: Lavsky, H. *The History of the Hebrew University of Jerusalem: A Period of Consolidation and Growth* (A), Jerusalem: The Hebrew University Magnes Press, 2005, p. 24. (Hebrew).

The initial years at the university were fraught with strife. The conflicts focused on the role professors would play in running the university and on how to assure high quality studies. Specifically, the professors opposed the autocratic governance of the university chancellor, Judah Leon Magnes. Albert Einstein, who was involved in the development of the university, had a close relationship with Fraenkel, which included an ongoing correspondence concerning university matters (in which Einstein relied on Fraenkel) and scientific matters. In the 1930s this conflict became so acrimonious that Einstein resigned from the university board of governors and suggested establishing a Jewish university in Lugano, Switzerland, instead. A letter from Chaim Weizmann to Fraenkel (in Holland) in 1933 states: "I cannot write you without referring to Einstein's action in publishing his criticisms of the University in the Press. At this present tragic juncture, this has been a very serious blow to the University, and especially to me and to those who, with me, have been doing their utmost to introduce the needed reform.... One of the main objects of my visit to Palestine/Land of Israel was to see what could be done in order to make the University into something really worthy of the name. My colleagues on the Constitution Committee and myself had given long hours of thought and discussion to the drawing up of a project, which I took with me to discuss with the Jerusalem authorities, and brought back with a real feeling that some progress had been made (though much remained to be done). But now Einstein's untimely action has rendered our task difficult, if not impossible—and that at a time when, in view of the tragic events in Germany, we needed every ounce of support we could get, and might reasonably have expected Einstein to be willing to throw in his lot with us. Moreover, I hear that he was even willing to lend his name to some hare-brained scheme for establishing a University for Jews somewhere in Europe—a scheme which, if it ever had the faintest chance of materializing, would never be anything but a scientific concentration camp for German refugees: an idea as absurd as it is harmful. It would not be possible to get a single penny for such a project—but nonetheless it might do a great deal of harm to the University.

I should be grateful for any information you may be able to pass on to me as to Einstein's present intentions and plans..."

What saved the university was an investigative committee organized by Weizmann. The committee published a secret report, following which, in 1935, the position of chancellor was eliminated; Magnes became president with more restricted functions, while academic functions were transferred to a rector, to be elected by the Senate. The philosopher Hugo Bergmann was elected first rector. As a result of these changes, the university was able to return to normal functioning.

Fraenkel served as the second rector of the Hebrew University (1938–1940). Those years included the Munich Crisis, Kristallnacht (November 1938), the beginning of World War II, with large portions of Europe conquered by the Nazis, and, in the Land of Israel, the Palestinian Arab Revolt (1936–1939), and the 1939 White Paper, which severely limited Jewish immigration to the Land of Israel. Those years brought about a sharp decline in the flow of students from Europe and donations from abroad were markedly reduced.

In various speeches, Fraenkel addressed ethical, religious, and practical topics that were central to the *yishuv*.[3] As university rector, he reminded his audience of the commandment, "Thou shalt not kill," emphasizing the importance of struggling against the Arabs and the British in an ethical manner. He also spoke about the importance of the Sabbath. One of his well-known lectures was about the commandment of "*Pru Urvu*", be fruitful and multiply. In response to the scant Jewish population in the Land of Israel and the declining birthrate, Fraenkel strongly advocated internal growth. He told students that it was their duty to have many children, adding, 'I am referring to those of you who are married'. As a result, after completing his tenure as rector and becoming pro-rector, he was humorously nicknamed 'pru-rector'.

His concern for the daily administration of the university continued after his tenure as rector ended. A letter from 1941 to the university administration details several inadequate processes, including the following, "mishandling and general neglect of university mail: There are occasions on which students and workers received their letters through the university mail already open. These included letters from government agencies or from the army, and in one case the letter was even stamped 'confidential'. Upon inquiry as to who opened the mail, the answer given was 'we do not know'. On other occasions, a worker would arrive at the financial secretary at around 11 A.M. on the day on which the salary was due. The clerk whose duty it was to deliver the check would respond, 'I have not had breakfast yet today, come back tomorrow. . ...'"

His concern for his students' well-being extended beyond their professional studies and careers to their physical welfare. He therefore promoted sports and physical education on campus and was the first chair of the university committee for physical education. He won approval for implementing sport as a required course for all students. He fought for a sports hall and a swimming pool on the University campus, maintaining that they were as important as lecture halls. At the time of his tenure as rector of the university, the *Vaad HaLeumi* (the National Council, the Jewish community's governing body prior to the State of Israel), instituted sport testing; in order to provide a personal example, he took the exams and surprised everyone, including himself, by winning a race.

Fraenkel was an outstanding teacher, who taught thousands of students, and raised a generation of excellent mathematicians, many of whom became professors in universities around the world.[4] "He also had a number of outstanding students who continued his scientific work.... His favorite lecturing method was to incorporate questions about potential solutions to problems which developed during the discussion. He awaited his students' answers, and through this method sought to

[3]Jewish community in the Land of Israel/Palestine at the time.

[4]Mazar, B., Poznanski, E. I. J, Rabin, M. The late Professor Abraham HaLevi Fraenkel, presented on 1st December ''65, Published by Magnes Press, Jerusalem 1965.

develop independent thinking and to discover through their successful answers, those students that excelled."[5]

Every seven years or so, Fraenkel used to give a seminar on "the Jewish calendar" for students in the departments of Talmud, Jewish history and mathematics. When students read their papers, Fraenkel would sit at the back of the hall, an unlit cigar between his lips and his eyes half closed. One could think he was dozing, but far from it. He never missed a beat. All of a sudden he would make a pertinent remark or ask an apt question related to what a student had just said.[6]

Since Fraenkel believed that true mathematical creativity occurred prior to age 25, he was always seeking out young gifted students. When he discovered an excellent student, he would cultivate and encourage him, not paying any attention to whether the student was interested in his (Fraenkel's) field of study, or had moved to another mathematical area. His help varied. He labored to get them awards and financial support, to connect them with appropriate teachers, and to organize international complementary studies for them after they had completed their studies in Jerusalem. A note at the National Library archive states that Fraenkel, who together with Professor Fekete, had guaranteed a student's debt, was requested to ensure payment after the student failed to pay up. No detail was too trifling for him; there was nothing he would not do to help his students. One of his greatest joys in the mid-50s was to find and cultivate exceptionally gifted children. After discovering one while lecturing at a kibbutz, he "immediately organized a specialized mathematics education for the child, found him teachers among the students at the university, made sure he was transferred to a good high school to prepare him for entry to the Hebrew University. One day Fraenkel said, 'I noticed that the boy was working too hard in his studies and did not look well. I therefore arranged swimming lessons for him, to strengthen his health'. Such was his interest in his students."[7] He strongly believed that promising young faculty members, rather than senior mathematicians, should be cultivated and lead the field and the department.

Fraenkel went on excursions around the country with his students, to the Judean Desert, as well as (before 1947) to Petra and other places in Eastern Transjordan, e.g., the summit of Mount Hermon. They rode horses, donkeys and camels. On a trip to Petra the local Sheikh argued that he wanted to buy one of the female students.

---

[5]Rabin, M. On the scientific work of A.H. Fraenkel, in: B. Mazar, B., E. I. J. Poznanski, and M. Rabin. The late Professor Abraham HaLevi Fraenkel, presented on 1st December 65, published by Magnes Press, Jerusalem, 1965.

[6]Source: Prof. Moshe Herr.

[7]Rabin, M. On the scientific work of A.H.Fraenkel. in: B. Mazar, B., E. I. J. Poznanski,, and M. Rabin. The late Professor Abraham HaLevi Fraenkel, presented on 1st December 65, published by Magnes Press, Jerusalem pages 12–16.

Fraenkel on trips with students

In 1946 Fraenkel received a letter from the University of Kiel inviting him to return to his position as professor at the university. The letter was in German with a translation into English. Fraenkel purposefully responded in Hebrew (despite German being his mother tongue) with a translation into English. He enumerated his reasons for declining the offer, including his early decision to move to the Hebrew

University, and also the humiliating dismissal letter he had received in April 1933 from the Nazi dean who had noted with scorn that Fraenkel always regarded himself as a Jew rather than a German. He added that he could not leave his people, his country and his university to live in a foreign land. At the end of his letter, he wrote, "I consider it objectively impossible, that a Jew, whoever he may be, would return to a country whose citizens, many actively and nearly all others with no opposition—were responsible for the annihilation of more than five million Jews, a third of my nation, in cruel torments that have not been seen for thousands of years. It would be intolerable to live among such a nation."[8] Related to his antipathy to Germany is a story recounted by Moshe Herr[9]: in the mid 1930s the German consul asked Fraenkel to see him at his office at the consulate. When Fraenkel arrived, the consul asked what German education he was providing for his children and how he fostered their love of the fatherland. Fraenkel replied, 'You should not worry, I provide my children with no German education and I do my best to instill in them a sense of loathing towards their motherland for the way it conducts itself.' In response the consul replied that according to instructions received from the German government, the monthly pension payments Fraenkel was receiving from the University of Kiel would be immediately stopped. (Since Kiel University had fired him after refusing to accept his resignation, he was entitled to a pension from them). After his dismissal in 1933 Fraenkel in fact never returned to Germany, in contrast to many other Jewish and Israeli scholars who accepted invitations to lecture there.[10]

His continued and complete rejection of Germany is further illustrated by another story told by Moshe Herr. In 1953, Fraenkel planned to travel to Switzerland. During the first 20 years after the foundation of the State of Israel, most other countries required visas of Israeli citizens. Switzerland was even more stringent, requiring as a pre-requisite to granting a visa, that the Israeli visitor have a second visa to a neighboring country (Italy, France, West Germany or Austria), so that if, for any reason, Switzerland chose to expel the visitor, he or she would be able to leave by train or car for one of these four countries. Moshe Herr, who was present at the time, remembers Fraenkel telling his wife that it was obvious that Germany and Austria were not an option, despite this being a mere formality.

In those days, in order to get a visa one had to go to the consulate in Tel Aviv and wait on line. One had to complete a form, present the additional visa, leave the passport at the consulate and return the next day to get the passport back. Since Fraenkel was not a resident of Tel Aviv, he hoped to receive the visa on the same day. Around two weeks earlier, the Hebrew University had held its Annual Student Day, which included, among others, a 60 meter race for faculty members over 50 years of age. Fraenkel, who was

[8]http://web.nli.org.il/sites/NLI/Hebrew/gallery/jewish/Pages/Jewish_Mathematicians.aspx

[9]Moshe David Herr is professor emeritus of Jewish history at the Hebrew University. As a student, he rented a room in Fraenkel's apartment for five years.

[10]For a discussion of Fraekel's approach to German colleagues after the war, see Remmert V.R. (2015). Forms of remigration: Émigré Jewish mathematicians and Germany in the immediate postwar period. *The Mathematical Intelligencer,* 37(1):30–40.

62 at the time, came first. The entire event was filmed and shown as part of the newsreel prior to movies in the cinemas. It turned out that a few days before Fraenkel's visit to the Swiss Consulate, the Consul General had gone to the movies and seen the newsreel. When he came to respond to Fraenkel's request, he called out to all those waiting: "Oh, there is the runner!" Inviting Fraenkel into his room, he immediately gave him the visa, saying it was an honor for him to meet such an esteemed scholar and athlete.

Upon his formal retirement in 1959, Fraenkel stopped his regular teaching at the Hebrew University, but continued to provide elective courses at both the Hebrew University and at Bar-Ilan University. He included new courses, such as a course on the history of mathematics.

Fraenkel received multiple awards. In 1946 he received an Honorary Doctorate in Humane Letters from Dropsie College in Philadelphia. In 1956 he received the Israel Prize, and in 1962 the Rothschild award. In November 1952 he was invited by Samuel Sambursky[11] on behalf of the prime minister to become a member of the Association for the Advancement of Science in Israel. He served as its chair for several years and in 1960 was among the founding members of the Israel National Academy of Sciences. When the President of the Academy, Martin Buber, notified him of his election and asked for his agreement, Fraenkel's response was:

> My dear friend Buber,
>
> I thank you for your letter of the 13 of Sivan[12] and announce my agreement to be a member of the Academy. However, I feel obligated to add that in my previous discussion with the prime minister over two years ago, I expressed my opinion that members should not be elected at an age over 65 years, and I have not changed my mind nor my reasoning in the principle. Though for you, my friend Buber, I always saw a reason for an exception because of your personality as much as for the special and wide field in which you worked and still work successfully. Since then other members were appointed or elected who are over the age of 65, and therefore some of my friends swayed me not to stand on principle on this point, and I have succumbed. This does not mean that I will not continue to fight for the election of specifically young members for the Academy, and especially for the natural sciences division.
>
> With blessing and friendship, I honor you,
>
> Avraham HaLevi Fraenkel[13]

In describing Fraenkel, many have noted that he was fearless in advancing his ideas and principles even when those were not shared by his superiors or by the opinions of the time. Of course, he collaborated with others to advance the projects he believed in, but when convinced of a specific stand, he stuck to it, despite others' opinions.

A number of Fraenkel's students have made important contributions to mathematics, including Abraham Robinson who succeeded him when he retired from his chair at the Hebrew University. Observing the success of his students around the world was a source of great satisfaction to him, and he was proud of teaching a whole

[11]Samuel Sambursky (1900–1990) was a professor of physics and natural sciences at the Hebrew University. http://www.jewishvirtuallibrary.org/jsource/judaica/ejud_0002_0017_0_17390.html

[12]Ninth Jewish month.

[13]Igeret, Journal of the Israel National Academy of Sciences, no. 20, May 2001, page 24.

generation of mathematicians who then taught at the Hebrew University, the Technion, the Weizmann Institute, in the Israeli army, at MIT, and in academic institutions around the world. In a 1968 interview with Eliyahu Shamir, Professor of Mathematics, Shamir notes, "...The strongest imprint was that of Professor Fraenkel. ... He had strange and amusing ways. He would invite students for an interview at his home at 6:30 in the morning. You can imagine a student who spent the previous night with a young woman until after midnight, and has to get up—after a few hours of sleep—at five, in order to manage to get to the professor's home at six thirty[14]—and God forbid they are late by even one minute. Until old age, Fraenkel would participate in swimming competitions with his students at Sports Day. Yet with all his exactitude, in important matters he was less formal than other professors. Unlike them, Fraenkel allowed his students to take the final examinations prior to the submission of their written projects, and did not stick to the letter of the law. When he "sniffed out" a gifted student, he would cultivate him personally and invite him to a Friday night dinner at his home (the professor was an observant Jew and Friday nights at his home were imbued with a distinct traditional atmosphere)."[15]

In his 2011 article 'Pure Science and Applied Science', Nobel Prize laureate Robert J. Aumann wrote: "Israel is the world's number 2 power in computer technology. The world's number 1 power is California, and number 2 is Israel. This is due to the school of thought that was founded by Professor Avraham HaLevy Fraenkel. Professor Fraenkel, the first chairman of the Mathematics Department at the Hebrew University, was one of the great contributors to "set theory", axiomatic set theory, in the 1930's through the 1960's. Set theory is a very abstruse branch of mathematics. I would not call it natural and easy to state; it is difficult, and it did seem at the time absolutely useless. Fraenkel raised a generation of scientists who became interested in what became the foundations of computer science. This is what has made Israel today the number 2 power in computer technology in the world".[16]

Publications

A partial bibliography of Fraenkel's writings is provided in this book. The bibliography reflects the wide scope of his writing and influence. Although the two foci of Fraenkel's identity, his interest in mathematics and related topics and his interest in Jewish heritage, are reflected in his writings, a wide range of other topics are represented

[14]6:30 A.M. should probably be 7:30 A.M., for Fraenkel only came back from the synagogue at 6:40 A.M. (on Mondays and Thursdays at 6:55 A.M.) and immediately sat down for breakfast.

[15]Ma'ariv, 1st of March, 1968.

[16]Aumann, R.J. Pure Science and Applied Science, Rambam Maimonides Med J. 2011 Jan; 2(1): e0017. Published online 2011 Jan 31. doi: 10.5041/RMMJ.10017 see http://www.ncbi.nlm.nih.gov/pmc/articles/PMC3678791/

including political activity and concern for education. He also wrote a number of obituaries for mathematicians, political figures, and people he knew personally.

Fraenkel studying a Jewish text

An examination of Fraenkel's bibliography reveals a constant flow of publications on mathematical topics. His writings were published every year (with very few exceptions) from 1908, when he was 17, until 1973, eight years after his death, with some publications appearing years later. He wrote several books on set theory, which were revised and reissued "*Abstract Set Theory*." (1953, reissued in 1966) and "*Set Theory and Logic*" (also published posthumously in 1966). The first edition of "*Foundations of Set Theory*" came out as a slim book of 150 pages, which then grew to 900 pages by the end of the 1950s. The final version was written with two of his students, Yehoshua Bar-Hillel and Azriel Levi. A review of this book describes it as: "A masterly survey of its field. It is lucid and concise on a technical level, it covers the historical ground admirably, and it gives a sensible account of the various philosophical positions associated with the development of the subject. This work is essential reading for any mathematician or philosopher—professional or prospective—".[17]

Committed to and interested in Jewish texts and traditions, Fraenkel wrote several articles relating to the history of the Jewish calendar, and also wrote about mathematical and astronomical research among the Jews. In 1960 he published *Jewish Mathematics and Astronomy*.

Another type of writing combining his knowledge of Jewish sources and his excursions around the land is exemplified in a short article he published in 1950, "The views and climate of the Land of Israel as a guide to understanding our ancient

[17]Bell, J.L., 1975, *The British Journal for the Philosophy of Science*, 26(2):165–170.

literature". A few lines will demonstrate his writing. "In Song of Songs chapter 1, verse 14, says the beloved woman, "My beloved is unto me as a cluster of henna in the vineyards of Ein-Gedi." The *midrashic*[18] interpretation of this verse is well known and found its place in the liturgical poetry.[19] But what is the simple meaning of this verse? What is the cluster of henna? And why should we search for it in the Judean desert? It seems to me that the best way to understand this verse is to take an excursion to Ein-Gedi... One can descend on foot through the desert from the mount of Hebron or take a boat for four hours from the north end of the Dead Sea. As you get closer to the sea, you see a green ribbon descending down the yellow desert to the Dead Sea—that is the stream going down from the spring, with fauna growing along its banks. You walk for about half an hour without particular difficulty, except for the heat that prevails here several hundred meters below sea level, and you reach the small pool adjacent to the waterfall. (Maybe you desire to wash in the cool pool water? And here the goats prevent you, as they wash in the spring and waterfall as they have done for thousands of years, for it is for them that the area is named Ein-Gedi—the Goats' Spring). You explore the area and find a huge contrast between the desert view that continues for a few hours walk in each direction and which allows only a few miserable salty plants to survive and the actual waterfall area, where there is, until now, fresh vegetation, even more copious in ancient times, as we know from the literature..."

As part of Fraenkel's commitment to provide public education, he edited the mathematics section of the Encyclopedia Hebraica and wrote many of the entries, such as 'Geometry' and 'Calendar'. He also wrote the 'Theory of Sets' entry in the Encyclopedia Britannica. This concern extended to the education of the young, for whom he published articles on the Jewish calendar in several magazines.[20]

---

[18]*Midrash* is an ancient manner of expounding the Bible which includes interpretations and stories that add to and fill in gaps in the original text.

[19]*Piyyut*.

[20]Fraenkel, A. H. (1943) על המבנה המתמטי של הלוח העברי מתוך דפים למתמטיקה ולפיסיקה לנוער המתלמד. טבת תש"ג חוברת ה'. [The mathematical structure of the Jewish calendar.] Dapim LeMatematica UlePhysica. [Pages in Mathematics and Physics for Youth]. Ch 1 and Ch 2. 5th booklet. (Hebrew).

Fraenkel, A. H. (1944) נוסחה המקשרת את הלוח העברי בלוח המושלימי. מתוך דפים למתמטיקה ולפיסיקה לנוער המתלמד . חוברת י' אייר תש"ד. [A formula relating the Jewish calendar to the Moslem calendar.] Dapim LeMatematica UlePhysica. [Pages in Mathematics and Physics for Youth]. (Hebrew) 4-6.

Fraenkel, A.H. (1948) על יסודות הלוח העברי. שיחה ראשונה. [The basis of the Hebrew Calendar. First conversation.]. *הצופה לילדים* כרך ב' חוברת 32-33 י, י"ג בניסן תש"ח, 22.4.48 עמוד 392 *HaTzofe for Children*. Vol 2 issue 32–33, 22 April 1948 p. 392 (Hebrew).

Fraenkel, A.H. (1948) על יסודות הלוח העברי. שיחה שנייה. [The basis of the Hebrew Calendar. Second conversation]. *הצופה לילדים* כרך ג', חוברת 1, תש"ט 3.9.1948 עמוד 14-15 *HaTzofe for Children*. Vol. 3, issue 1, 3rd September 1948, pp. 14–15. (Hebrew).

Fraenkel, A.H. (1949) על יסודות הלוח העברי. שיחה שלישית. [The basis of the Hebrew Calendar. Third conversation.] *הצופה לילדים* *HaTzofe for Children*. 27 January 1949 pp. 204–205 (Hebrew).

At the time of his death, Fraenkel had plans for further writing, including the next volumes of his autobiography as well as additional scientific writings.

Activity for the Public

As can be seen from the early chapters of his autobiography, and also from Fraenkel's will at the end of this chapter, Fraenkel was deeply influenced by his father's devotion to public good in general and to the Jewish community in particular. Fraenkel's commitment to those ideals is evident in the speech he gave at his own wedding (1920), "The life that I and my wife are together taking leave of, with a united heart full of happiness and hope, this life, neither of us views as our own lives, as lives provided to us as a present, and even not as lives granted only to the practice of the *mitzvot* [religious commandments] for us to enact towards our Creator. In no way can we separate our lives from the life of our nation, from the life of the whole Jewish congregation, which is currently facing troubles the like of which have nearly never previously occurred. With God's help may these current troubles produce redemption and salvation to the remnants of Israel. Therefore we confess and stress that both of us acknowledge and cherish the duty placed on us to do all that is in our power in order aid to maintain the existence of the remnant of the Nation of Israel according to the Torah and as much as possible on his own land which God had promised to the forefathers. We are committed to labor for this purpose throughout all of our lives. ..." The speech, composed and delivered in Hebrew (in Amsterdam), is concluded with "Next year in Jerusalem" and seems to foreshadow his future activities on behalf of the public, exemplified in the Center for Adult Education and in his work in secondary education. (See also Chap. 5, page 116).

Center for Adult (Popular) Education—Haskala La'Am

Fraenkel's idealistic commitment to education is exemplified in his multiple educational endeavors. Already in 1931, as a preface to his article "Beliefs and opinions in light of the natural sciences", a note appears that states that the article is based on lectures Fraenkel gave on this topic in Jerusalem, Tel Aviv, Haifa, Nahalal and Mikve-Israel. This type of work was epitomized in 1940 with his initiation and direction of the "Center for Adult (Popular) Education".

In his description of the history of the adult education project at the Hebrew University, Fraenkel explains his concept of the project, which at that time he had directed for about ten years. "One of the most important and difficult challenges faced by the State of Israel since its inception is in the field of education. Long after it emerges victorious from its military, political and economic campaigns, the battle on this front will continue. Decisions made there will determine the character of the State of Israel."[21] Fraenkel describes the need for educational institutions to shape the new nation within the context of people coming from varied backgrounds into a cultural melting pot.[22] This interest in adult education can be viewed against the European commitment to adult education in the period between the two World Wars. Adult education addressed diverse needs: teaching Hebrew to the many immigrants, providing basic education to those from countries where they did not receive it, and at the same time providing those with an advanced education with appropriate resources to meet their intellectual and cultural needs. In addition, Fraenkel considered adult education to be ingrained in Jewish tradition, where education does not end with school, but is a life-long pursuit.[23]

Already in the early years of the university, professors and lecturers gave lectures around the country, though the scope of this activity was limited. Fraenkel enjoyed recalling lectures that involved a ten-hour jeep ride at a stretch, riding a camel to Caesarea on another occasion, going on horseback to Manara and flying to Eilat.[24] However, the lectures did not satisfy the needs of the Jewish community in the Land of Israel, and in 1940 Fraenkel initiated a university committee for the people's education, which later became the Center for Adult (Popular) Education. Fraenkel chaired the Center from its inception in 1940 until 1957. In a 1944 article in the newspaper *Davar,*[25] he describes how in 1940, when contact between the Jewish community in the Land of Israel with the outside world was severed by the war, resulting in isolation of the *yishuv,*[26] the university decided to organize activities to promote learning in all parts of the Jewish community, including cities, towns, agricultural settlements and kibbutzim, even those located in remote parts of the country. The teaching method changed from sporadic lectures to regular courses and also included intensive summer courses. Courses included vocational training such as bookkeeping, as well as theoretical courses on a wide range of subjects, such as mathematics, physics, chemistry, biology, geology, various social sciences, humanities, and Judaic studies. Fraenkel's own lectures focused on topics such as modern mathematics, modern physics, Israel's climate with reference to the Bible, the Jewish calendar, and the Creation story in light of modern

---

[21] האוניברסיטה העברית בירושלים - כה שנה תרפ"ה – תש"י, עורך: חיים תורן, ירושלים תש"י, עמ' – 97 – 99

The Hebrew University in Jerusalem, 25 years, 1925–1950. H. Toren (Ed.), pp 97–99. Note: it is not clear whether Fraenkel wrote the chapter, but it is likely he was at least very involved in its preparation.

[22] *Ibid.*

[23] *Ibid.*

[24] אורות – לשאלות התרבות והחינוך קובץ ב', עורך נחום בנארי, תל – אביב 1954, עמ' 83 – 87

[25] Fraenkel, A.H. The University and Adult Education, *Davar* 10 February 1944.
א.ה. פרנקל, האוניברסיטה וחינוך מבוגרים, דבר 10 פברואר 1944.

[26] The Jewish communities in the Land of Israel/Palestine at the time.

physics.[27,28] The Center's activity aroused great public interest; new courses opened each year, attended by people from all parts of the country and from all stations in life.

Fraenkel teaching in Haifa 1944

Fraenkel describes the workers' motivation to study with pride and admiration. He writes that "in our settlements, hearts are unusually open to such education [adult education]. In Tel Aviv, the desire to study pure and abstract science is as great as that in cities of half a million people in Europe. There is scarcely a village which does not have workers with an academic education or with specific interests in scientific research."[29] He recounts that in various kibbutzim, lectures in the natural sciences generated discussions lasting for several hours; there were even cases where the Cultural Center coordinator had to turn out the lights to put an end to lively debates that continued beyond midnight. He also told of a kibbutz member who approached him a day after his talk, saying "last night you raised a

[27]Fraenkel-Bloch, R. Hazak! Hazak! Venitchazak חזק! חזק! ונתחזק! "Be strong, be strong and let us strengthen one another" A booklet in honor of the bar mitzvah of Joshua Feuerstein, 1995, p. 4.

[28]A list of lecture topics by different professors, including Fraenkel, is available in Adult Education: extension lectures and courses of the Hebrew University, Jerusalem 1943–1944. Jerusalem. Achva Coop. Printing Press. 19 pp. A copy exists in Bivliotheek Universiteit van Amsterdam. Item 0150000448.

[29]Fraenkel, A.H. The University and Adult Education. *Davar* 10 February 1944.

mathematical problem without providing the solution—I considered it during the night—and here is the solution."[30] He recounts a visit to the Land of Israel of Sir Arnold McNair, vice-chancellor of Liverpool University, who was invited by the Mandatory government to evaluate the Jewish educational system. Hearing about the topics of interest of village workers—topics such as logic, physics, mathematics, and chemistry—there was no end to his surprise and amazement.[31]

In a letter to Fraenkel, the head of the cultural center of the National Workers' Union, wrote: "You are first and foremost to any activity in the area of education for adult workers in its different venues, in the villages and in the city... The center for the nation's culture that you head, created and developed, has a great impact in enlightening the multitudes, to the benefit of the public and to the glory of the University, and serves as an important source of culture, which flows in different channels to all the far-flung places of the nation......in your wanderings to workers' localities in even the most remote and isolated corners, you have contributed greatly to the expansion of knowledge and wisdom, and you have made yourself friends and admirers among your listeners...."[32]

In pursuit of his interest in new models of adult education, Fraenkel toured several European countries in 1954, where he examined institutions and the history of adult education, searching for practices and innovations that could be exported to Israel.[33] Fraenkel believed that mutual enrichment could and should occur between all tiers of society and the university community. While acknowledging a world-wide trend towards adult education, Fraenkel enumerated several specific reasons for its requirement in Israel[34] "Firstly, education for the Jewish people does not traditionally end with graduation from yeshiva....Secondly, in many cases immigration to Palestine/ Israel involved radical vocational readjustment. Philosophers worked as bus drivers, poets as farmers.....Given their intellectual backgrounds, they had a great need for cultural stimulation.... where it was impossible for them to come to the University, the university would have to come to them. Thirdly....immigrants needed assistance to rapidly integrate into the community. They needed courses in the Hebrew language and in the history and geography of the country."

Similar in its aims to the National Education Center, Fraenkel's book 'Introduction to Mathematics'[35] sought to make mathematical concepts and knowledge

[30]Fraenkel, A. H. (1954) Issues of continuing education among workers. Orot – On Questions of Culture end Education. N. Banaari (Ed.) Tel-Aviv, pp. 84–87. (Hebrew).

[31]Ibid.

[32]Letter by H. Friedman, 1951, in the National Library of Israel ARC. 4* 1621.

[33]Israel National Library Archive, ARC. 4* 1621.

[34]Fraenkel, A.H. Adult Education. In: M. Spiegel (Ed.) The Hebrew University of Jerusalem 1925–1950, Semi-Jubilee Volume, 1950. Goldberg Press, LTD. Jerusalem. P. 111. Note: It seems like this chapter was written by Fraenkel, but the authorship is not clear.

[35]A.H. Fraenkel, Introduction to Mathematics. Magnes Press and Masada Publishing. 5 volumes. Published between 1942 and 1953. (Hebrew).

accessible to a wide audience. It also reinforced the Hebrew mathematical terminology which Fraenkel had been developing since 1929. Upon publication of the first volume (1942), Fraenkel received a letter from Prof. Joseph Klausner[36] saying: "I was pleased to hear that your book, 'Introduction to Mathematics' has been published. I have not seen the book, nor do I understand mathematics, but I am exceptionally pleased that our scientific literature merited a new book on a topic in which Hebrew writings were once rich, but where now there is close to nothing. I think that it is our sacred duty as professors at the first and only Hebrew University to prepare Hebrew books for the future in all the general subjects that are studied at the university, so that our students are not forced to draw upon foreign sources or on books written in foreign languages, which may diminish the status of our language in their eyes, and render their thought processes somewhat non-Hebrew. I congratulate you on the publication of your important book. Long may you continue to provide Hebrew books in all the scientific fields of your expertise…"

The book has become a classic Hebrew mathematical text. Its lasting achievement is illustrated by two anecdotes.[37] First, in 2013 the family received a request from scientists at Tel Aviv University to scan and upload his books, which are out of print and written in clear and simple Hebrew, in order to enable young scholars and other interested parties to learn from them.[38] The second example is a comment from an interview with Yehuda Meltzer, an editor and publisher who was interviewed for the *Ha'aretz* weekend magazine in December 2013.[39] He describes the development of publishing in Hebrew in Israel, and comments "Abraham Fraenkel, a mathematical genius, did not go to Princeton, but came here, established the Institute of Mathematics, and published books [*showing*] how to write mathematics in Hebrew. One cannot but be moved by this".

Commission for Secondary Education[40]

One of the initiatives that Fraenkel promoted in his first years at the Hebrew University was the Commission for Secondary Education. During the early years of the British Mandate in the Land of Israel (Palestine), secondary schools had almost no supervision. Many high schools requested that the Mandatory government recognize their matriculation certificates for transition to higher education.

---

[36]Historian and professor of Hebrew literature at the Hebrew University.

[37]See also the first pages of the introduction to this book by Prof. Magidor.

[38]This request was granted and the book is available at http://taharut.org/fr

[39]*Ha'aretz* Magazine 20 December 2013, p. 20.

[40]האוניברסיטה העברית בירושלים, כ"ה שנה תרפ"ה – תש"י, ישראל 1950, עמ' 83 – 85;
א' כהן, ההר והגבעה, האוניברסיטה העברית בירושלים בתקופת טרום העצמאות וראשית המדינה, תל אביב 2006, עמ' 257 – 263;
ARC. 4* 1621, Abraham Adolf Fraenkel Archive, Archives Department, National Library of Israel, Jerusalem
ב' ברשאי, "פרופסור אברהם הלוי פרנקל", דבר, 10 ביוני 1958, עמ' 3.

However, in the late 20s political conflicts emerged between the government of the British Mandate and the school principals; subsequently, the Jewish Agency's education department stopped sending students to take the British matriculation exams. Thus in 1930 the school principals approached the Hebrew University requesting that the university monitor the schools and examine the graduates. Fraenkel, aiming to help the *yishuv*,[41] especially in matters of education, and wishing to create a Hebrew educational autonomy in Israel, took the lead at the university. However, only after three years of discussion at the University and then at the Board of Governors, was the question of the supervision of secondary schools decided. In 1933 the Commission for Secondary Education was established; Fraenkel was its first chairman[42] and remained in that position until 1953. Representatives of the education department of the *Vaad HaLeumi*[43] joined the Commission only in the early 1940s.[44]

As part of the daily work of the Commission for Secondary Education of the Hebrew University, faculty members from the university visited high schools around the country and rated the level of studies. Students who attended Commission-approved schools could enter the university without taking an entry examination.

A number of pages documenting visits in March 1934 by various members of the Commission's academic staff have been preserved. They depict diverse schools from the religious Mizrachi teaching seminaries (which had gender segregation) and high schools (some segregated and some co-ed) as well as non-religious schools. Schools were located in Jerusalem and Tel Aviv. One visit involved the High School for Commerce (*Geulah*) in Tel Aviv, where the classes visited were *Mishnah*, *Talmud*, world history, Jewish history, and Bible. The visits usually lasted for half an hour to an hour. The comments: very good, "the interest of the (female) students and their knowledge are satisfactory", "discipline is not satisfactory. Students' knowledge, as much as can be ascertained in a short visit, is generally adequate", "cannot be evaluated because the students were new immigrants from Germany who had arrived in the past year, and therefore the results were generally inadequate", "inadequate, the proficiency in Bible (and especially in the Torah—Five Books of Moses) is under par for the students, but even for the teacher. The teacher talks too much and does not let the students think or answer on their own." As the chairman of the Commission for Secondary Education, Fraenkel also received reports from

---

[41]The Jewish settlement in the Land of Israel/Palestine at that time.

[42]כהן, אורי מוסדות ניהול האוניברסיטה העברית 1925–1948: תהפוכותיה של החתירה לאוטונומיה מוסדית מוחלטת. מתוך: תולדות האוניברסיטה העברית בירושלים התבססות וצמיחה א' עורכת חגית לבסקי. הוצאת מאגנס ירושלים 2005
Cohen, U. Governing Institutions of the Hebrew University 1925-1948, In: Lavsky, H. *The History of the Hebrew University of Jerusalem: A Period of Consolidation and Growth* (A), Jerusalem: The Hebrew University Magnes Press, 2005, p. 28.

[43]*HaVaad HaLeumi*—the National Council, the Jewish community's governing body prior to the State of Israel.

[44]א' כהן, ההר והגבעה, האוניברסיטה העברית בירושלים בתקופת טרום העצמאות וראשית המדינה, תל אביב 2006, עמ' 257 – 263;

other school inspectors. One such memo from May 1936, states "The Reali Gymnasium on Balfour Street, Tel Aviv: I have checked the sixth, seventh, and eighth grades. In addition, I was present, for a number of hours, at practicals in the chemistry laboratory of the sixth form. In comparison to the previous year, the situation has improved considerably, in my opinion, thanks to the teacher and the new pedagogical principal. However, in the seventh grade the level of knowledge does not meet even the minimal required standards. This conclusion concurs with the opinion of the teacher of the subject and of the pedagogical principal, who have demonstrated to me that in other subjects too this class fails to fulfil the study requirements. They are considering closing the class or not allowing the students to move up to the next grade. In my opinion, it is necessary that the university prevent any possibility of students from this class, in its current condition which cannot be fixed within a year, having the option of attending the university as regular students. As for the other classes, the situation is roughly similar to that in a number of schools long recognized by the university, despite the level not being satisfactory to the university teachers. In my opinion, in the case of chemistry, one has to recognize this institution, provided that in the coming years it fixes remaining deficiencies and will add the needed instrumentation. Here, as in other cases, the main difficulty is the lack of a curriculum that is accepted by all the schools for the study of chemistry, as such a curriculum has not been provided by the authorities." "The Reali School in Tahkemoni, Tel Aviv: I visited this institution and saw that the study of chemistry was discontinued a long time ago, because, according to the principal, it is impossible to find a teacher who will teach chemistry in addition to other subjects and from a religious perspective."

One of the anecdotes told about Fraenkel's inspector functions involved a visit to a religious high school for girls. The girls were scared when they heard he would attend a geography class. The teacher told them not to worry, that he would ask three questions about latitude and another three about longitude and that would be all. Fraenkel asked two questions about latitude and the answers were excellent, but for the third question the answers were not so good, because the girls had not noticed that he had moved onto longitude. The professor said: "the answers to my third question were excellent—but referred to my fourth question, which I will ask forthwith".[45]

The close supervision conducted by the Commission resulted in a series of amendments to the curriculum, changes in qualifications for the appointment of teachers and revisions in the organization of the matriculation examination. All these significantly raised the level of high school studies, as Fraenkel wrote: "teachers of the four highest grades were required to have an academic degree. The method of biblical instruction using textbooks which recounted biblical narratives in an abridged fashion [as opposed to study of the actual biblical text] was abolished. Instruction in Talmud was introduced for students in the humanity

[45]Based on an unidentified newspaper clip with the story attributed to Shlomo Stein of 5, Rav Avida Street, who was the teacher of this class.

track... The curriculum in the natural sciences was expanded with particular emphasis on laboratory work. The need for libraries was emphasized...".[46] Under Fraenkel's chairmanship, the Commission raised academic standards in secondary schools, added more required courses in Jewish studies, and increased experiential learning in the natural sciences. Fraenkel believed that education should not only impart knowledge, but more importantly, promote critical thinking and learning skills. He also believed that the humanities should be given the greatest weight in the high school curriculum—principally Jewish studies, which he saw as the foundation of Jewish education. He fought hard to establish Talmud as an obligatory subject for all high school students. During the 1950s the university Commission was replaced by the Ministry of Education of the new Israeli state.

Political Activity

As a fervent Zionist, Fraenkel was involved in political activity. He was a member of the elected national assembly (parliament) of the *yishuv*[47] during the British Mandate, and served as its vice chairman between 1944 and 1948.

Fraenkel started his political activities in the Agudah, and then turned to the Mizrachi movement. He was active in *Ha-Poel Ha-Mizrachi* (the Religious Labor Party) for 20 years. A delegate to the Twelfth Zionist Congress in Karlsbad [now Karlovy Vary] in September 1921, he continued as a member until the establishment of the State of Israel. He was a member of its elected council and presidential committee. In 1942 he co-authored a public letter calling for support of Elitzur, a religious movement which, under a front of being a sports organization, recruited young people to the *Hagana*, the organization for the self-defense of Jews during the British Mandate. In his letter explaining his resignation from party activity (in 1950), he expressed reservations regarding some of the policies which, in his opinion, did not befit a workers' movement, and further explained, "But the main reason for my departure is based on the inner chaos that has spread in the previous two years in *Ha-Poel Ha-Mizrachi*, through constant hustle among the subgroups and through personal conflicts of interest and activity of secret groupings. This harassment has weakened the focus on the goals of the party of "Torah and Labor" and sometimes led to activities that directly contradicted the interests of a religious workers' party." In 1961, he joined the Liberal party because of its intention to promote a system of national health insurance and to improve the education system

---

[46] א' כהן, ההר והגבעה, האוניברסיטה העברית בירושלים בתקופת טרום העצמאות וראשית המדינה, תל אביב 2006, עמ' 262 – 263

[47] The Jewish population of the Land of Israel/Palestine.

and the economy,[48,49] Eventually he left all the parties in which he had been involved at one time or another. In an interview on his 70th birthday,[50] he is described as being disappointed with all the parties. He was upset about the lack of national health insurance and about irresponsible monetary policies, in which the value of the Israeli currency was artificially controlled.[51] He also criticized policies relating to high school teachers which provided no incentive for such teachers to undertake academic studies. In his encounters with the ministers of education, Fraenkel always insisted that the best teachers be sent to the poorest schools and neighborhoods. The scope of his educational activities and political efforts derived from his grasp of the needs of the young state and his vision of the state as a Jewish spiritual center. This deep conviction and far-reaching vision are especially remarkable when seen in the context of his demanding academic life.

In this same interview (1961), he said, "if there were an atomic war, it would not be worth surviving it, and therefore there is no point to all the efforts devoted in Europe and America to assuring survival of body and properties." However, he thought there was no imminent danger of atomic war. The article concluded with his saying "It is interesting to remember the days in which Magnes argued that two mathematics professors are too many for the faculty."

Fraenkel was active in the voluntary organization *B'nai Brith*, and served as its "Great President" (i.e., president of all 43 lodges) in 1949–1952. He saw this role as an opportunity to promote his educational ideals and through his *B'nai Brith* connections he hoped to advance the Hillel Jewish students' organization at the Hebrew University. In 1952, he told the annual meeting of the *B'nai Brith* national executive committee in Washington DC: "*B'nai Brith* is a great moral force in Israel today because it supplies the only platform where people of different political parties can meet and discuss major problems in a spirit of unity."[52] A eulogy by Ben-Zion Kauders describes some of Fraenkel's accomplishments at *B'nai Brith*. "Fraenkel was elected at a difficult period. A quarter of the lodges were inactive, a quarter showed no activity, there was no secretariat, and the cashbox was empty." Fraenkel organized the work, and during his presidency the dormant lodges woke-up, and 14 new lodges were established, some of which were considered the best lodges. He developed lodges for young *B'nai Brith* members.[53]

[48]An interview with Fraenkel: מדוע הצטרפתי למפלגה הליברלית? Why did I join the Liberal party? Ha'Boker 26 May 1961. p 3. (Hebrew).

[49]מצטרפים למפלגה הליברלית Joining the Liberal party. *Maariv*, 7 July 1961, page 12.

[50]כשהפרופסור לא היה פרופסור [When the "professor" had not yet been a professor] *Pi Ha'Aton*. 6th December 1961. (Hebrew).

[51]Both issues were addressed in future decades.

[52]http://www.jta.org/1952/02/06/archive/bnai-brith-activities-in-israel-and-britain-reviewed.

[53]Kauders, B. Professor Avraham HaLevi Fraenkel. Yedion *B'nai Brith*, The Great Lodge no. 14, May 1966, page 71.

Committees

In 1958 Fraenkel resigned from a committee to assure integrity within the civil service. In his letter of resignation,[54] he described the purpose of the committee and its members, and explained his reasons for resigning:

"During the First and Second Knesset (Israeli Parliament), the public committee worked at an orderly pace, and there were no instances of ministers' opposition. This state of affairs changed almost immediately after the election of the third Knesset, especially on the part of two ministries, those of development and health. At the beginning of 1956, about half a year after the election of the third Knesset, I commented on cases that were in conflict with the committee's role, and during the last two years, the committee discussed, in several of its meetings, more than a dozen such cases in those two ministries. It is sufficient to mention here the appointment of Mr. Preminger that took place, as did several others—despite the categorical opposition of the State Service Commissioner.

After the protests of the committee and of several of its members came to naught, I asked you to convey my resignation to the Prime Minister. In your letter of the 17th of November 1957, you communicated to me and to Judge Vitkon [of the Supreme Court] your discussion with the Prime Minister and his request not to disperse the committee at this point. I responded to the Prime Minister on the 2nd of November 1957 that I conceded out of a feeling of respect and appreciation for the Prime Minister, and that I would refrain from resigning now, but with an explicit condition that such cases would not reoccur.

A short time thereafter, the minister of health appointed Ms. Haya Tzipkin, director of the nursing department, and in the meeting of the public committee on the 2nd of December 1957, the State Service Commissioner told us that he had told the minister of health that he would not be able to approve this appointment. After discussion in various forums, including with the Prime Minister and the Minister of Justice, this affair also ended with the approval of the appointment, in April 1958.

This, I feel—in keeping with my message to the Prime Minister—is the straw that breaks the camel's back. I cannot continue, by my association with the public committee, to take responsibility for the public perception that the appointment of civil servants is conducted with no regard for their party standing.

I thank you for the positive regard you manifested as the committee's chairperson and ask you to inform the Prime Minister [or the appropriate clerk in his office] of the contents of this letter."

[54]Israel National Library Archive ARC. 4* 1621.

Private Life

Religious Life

While Fraenkel led a strictly observant Jewish life, his religious outlook was complex and possibly evolved over time. This complexity can be demonstrated by two close witness testimonies. One close relative described seeing him praying alone, and said she had never observed such religious devotion in prayer. In contrast, another close relative believed that he was not religious in his beliefs and that for that reason he never discussed religious belief and doctrine with his children. It was her understanding that he maintained a religious life because that was what he had promised his father on the latter's deathbed. His ethical will, translated below, attests to the importance of religion in his life, though that position is somewhat attenuated in the final comments to the will. In a 1939 letter[55] to the Chairman of the Jewish Agency, David Ben Gurion,[56] Fraenkel wrote: "Yesterday noon, as I was walking to Rehavia,[57] several people approached me complaining that in the courtyard of the Jewish Agency, the gardener was watering plants with a hose-pipe.[58] I went there and told the gardener that it was the Sabbath. But, instead of stopping his work, he responded with dirty and insolent words." Fraenkel then requested that the head of the Jewish Agency handle the matter as he (Fraenkel) had persuaded others not to bring the issue to the newspaper, promising them that he himself would inform the heads of the Jewish Agency of the matter, since they were surely unaware of this desecration.

Another source which may provide some insight into his religious beliefs is his letter to his daughter Tirza after she informed him of her decision to cease Orthodox practice. "Every independent thinking young man or woman around the age of 18–30, in the previous and the current generation (and maybe also in previous generations in Western Europe), has gone through a crisis of belief; some remained caught up in the crisis and came to far-reaching decisions, some hovered between the two options, some recovered as they approached 30 (there might be a physiological law here) and found, like me, that as difficult as it is to believe, it is even more difficult to develop a life-vision and to observe the world without belief. After the First World War, intellectuals from the liberal wing of the *Mizrachi* movement in Berlin formed a group whose heads were [Oscar] Wolfsberg and, I think, Yeshayahu Leibowitz. Their motto was 'orthoprax not orthodox'; that is, we will keep the *mitzvot*, the religious laws, for national reasons despite our lack of faith. The later development of those people, of whom Leibowitz was the most famous, is known".

Fraenkel attended the daily synagogue prayers throughout his life. For his daily prayers he would go to the Yeshurun synagogue near his home; on Friday evenings

[55] Israel National Library Archive ARC. 4* 1621.

[56] He later became prime minister of Israel.

[57] A residential neighborhood close to where he lived in Jerusalem.

[58] According to Jewish law, watering plants is not allowed on the Sabbath.

he liked to rotate among different synagogues of Jews from different parts of the world—Yemen, Bukhara, Urfa, Kurdistan, Iran, North Africa, Spain, and Europe. He often took his Friday night guests on a tour of the different synagogues, and, since Jews from different origins start the Sabbath prayer on Friday nights at different times, he could show different customs and not miss the prayer. On Saturday mornings he would attend the early prayer at the Kehal Hasidim synagogue in Shaarei Hessed, where he officiated as the public reader of the weekly portion of the Torah, excelling in the exact reading of the Torah. For many years he participated in a weekly Sabbath Talmud class taught by Rabbi Professor Simha Assaf. And on the Day of Atonement he would go to the small Binyan Zion synagogue, where they followed the customs, practices and melodies of the Rhineland (south-west Germany), which were familiar to him from his childhood. The main reason for this preference was the practice in this synagogue of praying *selihot*[59] during all five prayers of the day; he loved *selihot*.

His Passover *Seder* (service conducted at home) nights included detailed learned conversations. He always discussed the significance of miracles experienced by the Jewish people during the late nineteenth and first half of the twentieth century. These insights he also summarized in an article.[60] The meal began around midnight or even later. It ended with the singing of psalms and *piyyutim* (liturgical poems). Having gone on for about eight hours, the night would finish at four in the morning.

When several *yeshiva* students who had fled to Japan during the Second World War wanted to know on which day the fast of *Yom Kippur*[61] fell (i.e., where the International Date Line was from a *halachic* perspective), they approached the Hazon Ish (Rabbi Y. Karelitz)[62] and the Hasidic Rabbi of Gur. The Hazon Ish relied for his response on the writings of Rav Zerahia Halevi, otherwise known as the "Baal ha-Maor" (author of the Book of Light) who lived in Provence in the twelfth century. He responded that the Date Line for *Yom Kippur* observance was at the end of Asia (implying that Japan has the same day for *Yom Kippur* as the rest of Asia). The Rabbi of Gur asked Rabbi Herzog[63] who passed the question to Fraenkel. Fraenkel, who relied on the travels of Marco Polo, claimed that the Baal ha-Maor could not have known of the existence of Japan, and that the Date Line should therefore be east of Japan rather than west (implying that the day for *Yom Kippur* in Japan was not the same as in Asia). The students thus received conflicting instructions, and some ended up fasting for two days.[64]

[59]Jewish penitential poems and prayers, especially those said in the period leading up to the High Holidays and on Fast Days.

[60]פרנקל פרופ׳ אברהם הלוי במסות באותות ובמופתים ובמלחמה ובמוראים גדולים המעיין א, ב (תשרי תשי״ד) 7-1.
With Tribulations, signs and wonders, and in war and in great terror. *HaMaayan* 2 (1), 7–11, 1953.

[61]Day of Atonement.

[62]Rabbi Karelitz (1878–1953), named for his book, the *Chazon Ish*, was one of the leaders of ultra-orthodox Jews.

[63]Rabbi Isaac Yitzhak Halevi Herzog (1888–1959) was the *Ashkenazi* Chief Rabbi from 1936 until 1959, first of the *yishuv* and then of the State of Israel (from 1948).

[64]Igeret, Journal of the Israel National Academy of Sciences, no. 20, May 2001, page 23. http://www.academy.ac.il/data/egeret/33/EgeretArticles/20p22.pdf

Fraenkel wrote and lectured on the lack of conflict between science and religion. His 1931 article "Beliefs and opinions in light of the natural sciences"[65] makes two main points. The first is the fallibility of the natural sciences, where he provides multiple examples from physics and chemistry, demonstrating how supposedly unshakable universally accepted notions and theories changed with developments in science, thus suggesting that our current scientific concepts are subject to revision and modification. The second point is that the Torah did, and had to, "speak in the language of humans", and thus does not offer a scientific depiction of the universe. In other words, he maintained that any apparent contradiction between Jewish holy literature and scientific notions stems from a simplistic reading of the religious sources and misses their intention to teach fundamental ideas rather than science.

Fraenkel resigned from his 20-year membership in the religious party *Ha-Poel Ha-Mizrachi*, which supported religious rule as part of state law. In Chap. 5 he writes, "After the founding of the State of Israel, I broke off with them .... as I did not think "religious" parties in an independent state were in keeping with the times". According to Moshe Herr and others,[66] Fraenkel believed in the separation of religion and state in Israel.

The notion that his religious stance may have also gone through a certain transformation is reflected in his wills. In 1940 he wrote, "I thank the Benefactor Almighty, that has bestowed his grace on me, that has bestowed on me rescue, deliverance, success, blessing and salvation, and above all, that he favored me with a helpmate, with life in the Land of Israel, with educating my sons and daughters to Torah and *mitzvot*[67] (and I pray and hope that they will continue to maintain a religious life in the future)", while in 1963 he adds the following comment, "But I want to say explicitly, that each one of our sons and daughters is, in my humble opinion, entitled to arrange his life to the best of his conscience. If, on the one hand, I enjoy the fact that Benjamin and Aviezri continue to study Talmud, on the other hand, I do not disapprove of Tirza because she and Jacob educate their daughters in a different way than Benjamin and Judith. I am deeply impressed by the seriousness and the loyalty with which Tirza and Jacob try to solve their particular and difficult educational challenges." Whatever his inner convictions, Fraenkel's conduct was strictly religious, yet with a modern, scientific, and open point of view.

[65]An analysis of his philosophy of religion and a translation of this article into English can be found in Zelcer, M. (2011) "A.A. Fraenkel's Philosophy of Religion: A Translation of "Beliefs and Opinions in Light of the Natural Sciences". Hakirah, Flatbush Journal of Jewish Law and Thought, Book 12, 209–232. http://www.hakirah.org/Vol%2012%20ZelcerM.pdf

[66]Also mentioned in מצטרפים למפלגה הליברלית Joining the Liberal party. *Maariv*, 7 July 1961, page 12.

[67]Study of Jewish texts and keeping of Jewish law in daily life.

Family and Social Life

In the 1930s Fraenkel devoted considerable time, effort and personal expenditure to helping people leave Germany, and then Europe. Fraenkel was the family pioneer in emigrating to Israel. He influenced his entire extended family to follow, and indeed his mother, sisters, brother, and brother in law all moved and settled in the same neighborhood during the 1930s. Every Saturday morning the Fraenkel siblings and their spouses, their widowed brother-in-law, Samson Breuer, and his second wife, would go to see their mother and get her blessing. Fraenkel also tried to convince the wider family, including his uncles, aunts and cousins to leave Europe. In persuading them to come, not only did Fraenkel present the Land of Israel (at that time named Palestine) as a haven in a time of danger, but also as a politically and spiritually viable homeland. He was the first family member to make trips throughout the land and to find beauty in the desert's barren landscape, despite its sharp contrast to his native verdant Bavaria and the snow-topped Alps.

In addition to facilitating the safe extrication of his extended family from Europe, Fraenkel was also involved in the rescue of many academics. Extant letters attest to his efforts to bring professors from Germany to the Hebrew University. Less well documented are his efforts to bring others to Israel. At that time, a sum of a 1000 British pounds was required as a guarantee in order to get a certificate for entry into Mandatory Palestine.[68] Fraenkel offered his personal funds as loans to strangers to enable them to enter the country. Since the loans were given on a basis of trust, no records were kept. Given that Fraenkel himself was not given to self-praise, he never mentioned those loans, but family members were told about them through chance meetings with people who attested, first hand, that he had saved their lives.

While Fraenkel was able to bring most of his family to Israel, he was not always successful. The most notable exception on his side of the family pertains to his mother's sister, Therese Loewy, widow of his uncle, the mathematician Alfred Loewy,[69] who had died in 1935. Her only daughter had died as a baby and she refused to leave Germany. Her entire identity was bound up with the cultural and intellectual life of the university. She felt supported by her many gentile friends and probably assumed that no one would dare harm the widow of a well-known professor. She committed suicide on the day the Gestapo was scheduled to come and take her away.[70]

Family members of Fraenkel's wife, Wilma, were even less fortunate. Wilma's mother, Emma Prins, refused to leave Holland, home to the graves of the family ancestors, and Wilma's sister, Meta Posen, remained there together with her

---

[68] Starting in 1920, the British Mandate government required an immigration authorization ('certificate') to enter the Land of Israel. The number of certificates was limited, and only those with independent means, defined as having over 1000 British pounds, could obtain a certificate outside of the quota.

[69] http://www-history.mcs.st-and.ac.uk/Biographies/Loewy.html

[70] See also Chap. 1, page 11.

husband, Eli, and son Bob. They were taken to Westerbork concentration camp in Holland, from where they were deported to Bergen Belsen. Their fates are described in a July 1945 letter by Bob's wife, the only survivor, to her brother-in-law, Phillip Posen, who had been sent to England before the war:

> "I believe you already know what has happened to your parents in Bergen Belsen.[71] It must have been terrible for you to hear that. The only thing I can tell you is that they were all very brave, especially your mother, who was a wonderful woman, everyone in the camp loved her because she was friendly to all. First of all, your grandmother died on the 3rd of June. The next was Bob who died of exhaustion; he suffered like no-one else from hunger. He could work very well but only if he had something to eat and he had to work very hard while the food wasn't enough and very bad. So he fell ill. The first time he recovered, but the second time he got dysentery and died on the 4th of October 1944. At first I couldn't grasp he was dead. Bob was always healthy, but I had to, you know; in the camp it was like that, you didn't have time to think, and that was good, because otherwise you would go mad.
>
> You probably know that your father was diabetic and had a wound on his foot that got worse and worse until he died, on the 3rd of January. Now your mother had lost her mother, son and husband, but she didn't lose her vitality and wanted to live to see you again, but she didn't have the strength and was already exhausted after your father died. The last night she was with him and made him imagine that he was in Frankfurt again and all was good and happy. I can't tell you how beautiful that night was for your father, he forgot all the misery of the last few months and totally lost his bitterness, because he was very bitter when the Germans murdered his son and let him starve. Your mother was really wonderful and so brave, but she also fell ill, or more exactly, she could not stand up anymore and a day before we left Bergen Belsen, she died. It was on the 8th of April. The cause of her death was also exhaustion...."

In addition to these family members, most of the social networks of Fraenkel and his wife were lost in the Holocaust, or, at best, struggled with displacement around the world. This is exemplified in a letter that Wilma, Fraenkel's wife, wrote to her siblings in 1944 telling them what she had heard from a few former inmates of Bergen Belsen who had been exchanged with Germans and arrived in Jerusalem where she met them: "Tante [aunt] Betti and Cobi and the family are still in Westerbork (the Dutch concentration camp), her son in Poland [they were probably killed in Auschwitz, but Wilma did not know that at the time]. Julie Konijn and her husband in Celle [i.e. Bergen Belsen], their children in Poland. Mietje Kan, the mother of Roosje, died. Rabbi Coppenhagen died here a short while after his arrival; my friend, Rabbi de Vries, died in Celle and also Ruben Cohn...."

In 1942, it looked as if the Germans were winning in Europe and were conquering North Africa in an attempt to reach the Middle East as well. The lives of the refugees from Europe in the Land of Israel seemed to be threatened again. The victory of General Montgomery over Field Marshal Rommel at El Alamein saved the Jewish community there. Fraenkel experienced this victory as a miracle, and thus the family's rescue from Europe and the victory at El Alamein became a feature of every Passover *Seder* as a personal redemption story.

Fraenkel had relationships with many of the leaders and intellectuals of the country at the time, as evidenced in the notes and messages he left. He had a special

---

[71] The Bergen-Belsen Nazi concentration camp was near the town of Celle in Germany.

relationship with Albert Einstein with whom he corresponded on matters that included, but were by no means limited to, mathematics and university administration. He was a friend of Shmuel Yoseph Agnon, who received the Nobel Prize for literature after Fraenkel's death. Another close friendship was with the artist Herman Struck.

Beit HaMaalot—picture taken in 2014

Home

After renting an apartment in the then new Jerusalem neighborhood of Rehavia, Fraenkel was involved in the innovative endeavor of building an apartment house a little outside Rehavia around 1935. The new building, named *Beit HaMaalot* or "House of Ascending Stairs",[72] is on the slope of a street perpendicular to the main

[72]The name has multiple meanings and connotations. It refers to the fact that the building looks like stairs of buildings and the balconies are arranged step-wise. It connotes the Psalms, chapters (120–134) starting with "A song of ascents", which uses the same name, 'Maalot' for ascents. It also means advantages, thus, House of Advantages. See http://www.jerusalem.muni.il/jer_sys/picture/atarim/site_form_atar.asp?site_id=11576&pic_cat=1&icon_cat=5&york_cat=8&type_id=28;

King George Street. It has seven entrances and five or more floors in Bauhaus style. The first building in Jerusalem to be equipped with an elevator, it became one of the town's attractions. The luxurious features included central heating as well as laundry rooms. Originally, a synagogue was constructed in the basement and a *mikve*, ritual bath, on the roof; however those were never used.[73] It also had an original (non overlapping) balcony design to allow for the building of *Sukkot*.[74] Many well-known Jerusalemites lived in the building. Fraenkel, who was deeply involved in the planning of the building, was also the head of the original residents' committee of the new cooperative endeavor and in the 50s still chaired the tenants' committee of two of the entrances. Like several of their neighbors, the Fraenkels bought a double apartment, on the fifth floor, with balconies overlooking the Old City of Jerusalem, Mount Scopus, the Mount of Olives and the Moab mountains. In addition to the living room with furniture brought from Germany, there was a study for Fraenkel. Like some of the rooms in Hensel's[75] home, the office had double doors separated by a few centimeters, to assure soundproofing. One of the walls was taken up by a bookcase. In later years, he enjoyed sitting in a green reclining chair from which he would write leaning on a wooden board. There were four bedrooms: one for Fraenkel and his wife and the other three for their four children and, over the years, a number of other persons. One of them, Fraenkel's nephew, was a son of his deceased sister, Else Breuer, who shared a room with one of the Fraenkel sons, then with the other son, for over six years. A daughter of a cousin, who came to study in a Jerusalem seminary, also lived there for a while. Mrs. Fraenkel's unmarried half aunt, Estella (daughter of her grandfather, Liepman Philip Prins, by his second wife), who escaped from Germany, resided there from the late 30s until her death in the late 60s. Estella, a year younger than Wilma Fraenkel, was welcomed into the family. She had no profession or means of supporting herself and never mastered Hebrew. Since Fraenkel's income between the later years of World War II and the mid-50s was insufficient for the household needs, one of the rooms was rented to students for several years. One such student was Isadore Twersky[76] and later Moshe Herr who rented the room (with board) for five successive years during his studies at the Hebrew University. For a while two rooms were rented. Farha, a single immigrant from Iraq, with no close family, rented a room for some time. While some of these occupants came there through family connections, they all participated in family meals and thus renters and family forged strong bonds which in most cases lasted a lifetime.

While Fraenkel saw family as the center of Jewish life in general, and of his own in particular, he was a strict father, who believed children should not be spoiled. His

---

[73]*Kol Ha'ir*, 31.7.81, p. 25.

[74]Living spaces, like small huts built during the Jewish festival of *Sukkot*, which, according to Jewish law cannot be built under a permanent roof.

[75]Kurt Hensel was Fraenkel's professor in whose villa in Marburg the Fraenkel family lived when Fraenkel was a professor there. See Chap. 5, page 116.

[76]Isadore Twersky (1930–1997) was later an Orthodox Rabbi and professor of Hebrew literature and philosophy at Harvard University.

wife, Wilma, who took care of the home and the children, was also involved in community work. During the Second World War, she set up a hostel for homeless young women, prostitutes and others. Some of the women gave birth at that hostel, and all were helped to reintegrate into society. In 1953, Wilma started an organization named *Keren Yaldenu* (Our Child's Fund), dedicated to the protection, welfare and supplementary education of socially and culturally disadvantaged children through the establishment and operation of after-school youth centers and clubs around the country. Children could attend after-school classes in a large variety of subjects (mathematics, physics, chemistry, English, French, electronics, sewing, knitting, weaving, woodwork, sports, crafts, electronics, mechanics, music appreciation, and later computers). The children received lunch and clothes as needed. In 1968 the project included 32 centers with a weekly attendance of 10,000 children.[77]

Fraenkel and family in the early 40s: From right: Fraenkel's wife—Wilma, Fraenkel's nephew—Vivian Prins (British soldier in WWII), standing—Rahel Fraenkel, sitting—Aviezri Fraenkel, Benjamin Fraenkel (British soldier in WWII), Tirza Fraenkel, and Fraenkel himself

Personal Characteristics and Hobbies

One of Fraenkel's hobbies was meteorology, which he had learned as a soldier in World War I. Several meteorological instruments hung on the walls of his study. His weather predictions were mostly very accurate.

[77]The Jerusalem Post Magazine – Family page, 15 Nov 1968.

Fraenkel was very strict about punctuality and order, and about the exact and correct use of the Hebrew language. In a lecture at *B'nai Brith* in 1944, he said: "Lack of punctuality and waste of time is a theft... theft of time is a damage that cannot be paid back.... you are invited to a gathering, a meeting or a party for a specific hour. You bought a ticket for a play in the theater or to an inter-city taxi ride at a specific time; if you arrive at the appointed time, you have sinned and will receive a punishment. Those who are late by half an hour or an hour, will be waited for and the show's start postponed for them... Of course, based on our notions, one has to punish the punctual and reward the late. We have a concept of 'Jewish punctuality' rather than of 'German punctuality'".[78]

Fraenkel appreciated, even adored, the literary beauty of the Hebrew Bible, the prayer book and of several *piyyutim*, especially those of medieval German Jewish poets like Shimon bar Yitzhak. Having studied in a classical *Gymnasium* in Munich, Fraenkel had a thorough knowledge of ancient Greek and Latin. He used to say: "One who does not know Latin does not know what culture is, but one who does not know Greek does not know what pleasure is". He often quoted Greek literature, especially Homer, whose works he admired. Once, during a discussion of literature, someone mentioned Goethe's oeuvre and alluded to his epic poem, *Hermann und Dorothea,* which is written in Homeric hexameter. Fraenkel, who was very familiar with the works of Goethe and Schiller, which he had studied at school, was shocked and said: "How could you even dream of comparing the coarseness and the roughness of the German language with the beauty and gentleness of Greek?" He nonetheless appreciated the novels of Thomas Mann as well as the dramatic works of the Austrian Jewish author Richard Beer-Hofmann [see chapter 1, page 16]. However, he did not show much interest in modern or contemporary literature. Suffering from insomnia for many years, he used to turn in quite early (at 8:15 most nights of the week) and, having swallowed a (sometimes double) dose of a barbiturate, would take with him to bed a detective or other popular novel that would help him fall asleep (he usually awoke by two in the morning). He was not much interested in fine arts nor in music, the only exception being Wagner's opera *Tannhäuser,* which he had liked to hear as a boy. A few years before his death his wife bought him a record of the full opera.[79]

[78] *Ha'aretz, HaShavua* magazine, 18 Nov 2011, p. 21, and Igeret, Journal of the Israel National Academy of Sciences, no. 20, May 2001, page 25, and a special reprint from *B'nai Brith* newsletter, 5, 1944. נמוסים בישוב. מתוך דברים שנאמרו ע"י האח פרופ' א"ה פרנקל באספת לשכת "ירושלים" סיון תש"ד. תדפיס מיוחד מתוך "ידיעות בני ברית" חוברת ה', תמוז תש"ד. יוצא לאור ע"י לשכת הגליל הגדולה מס' י"ד, ירושלים אב תש"ד.

[79] Source: Moshe Herr.

Folklore

Even at this point, over 50 years after his death, anecdotes about Fraenkel are still told in Jerusalem. Those anecdotes, many of which are fables, usually demonstrate his logic and exactitude. One such story describes him going on the bus to the university on Mount Scopus, when a student sitting next to him asked, 'would you please shut the window, it is cold outside', to which Fraenkel responded, 'and suppose I close the window, will it be warm outside'?[80]

Fraenkel liked swimming and when he happened to spend a day or two near the seashore he would swim, even in winter. Once, when he was on the verge of doing so, a lady said to him: "Sir, but it is raining". He retorted: "So what of it?" She then said: "You will get wet", to which he said: "And if I enter the sea when it does not rain, would I remain dry?"[81]

It is recounted that Fraenkel spoke of the university as having two dictators in its initial years, Magnes, and the driver of the number 9 bus going from town to the university. It was the driver's mood that dictated who would ride the bus and who would get there on foot.

Another anecdote was that "Fraenkel always carried two watches so as to be sure what the time was. One day we visited him and asked if he would devote five minutes of his time if he was not busy. That would be fine, answered Fraenkel. We started talking. After five minutes, Fraenkel interrupted us in the middle of a sentence: 'I have to tell you that five minutes are over. I did not say I was busy, but you asked for five minutes'. 'We request another five minutes,' we said. 'That is fine' Fraenkel replied".[82] As a matter of fact, he often had three watches, one on each wrist and one in his waistcoat pocket, so that if two of them contradicted each other, the third would determine the right time.[83]

Another story concerned his many lectures on science around the country. Fraenkel arrived to give a lecture at Ein Harod,[84] a kibbutz in the north of the country. The lecture was supposed to start at 8.30 in the evening, but at that time there was no one in the lecture hall. People arrived at 9.00 and at 9.15 the lecture started. Fraenkel opened by saying, "Ladies and gentlemen, this lecture was supposed to start at 8.30 and to end at 9.10. It ended five minutes ago". He did not deliver the lecture.[85]

---

[80]Dyokan, Makor Rishon, 18 June 2004; versions of the same story appeared in *Ha'aretz, HaShavua* magazine, 18 Nov 2011, p. 21, and in Igeret, Journal of the Israel National Academy of Sciences, no. 20, May 2001, p. 23.

[81]Recounted by Moshe Herr as having been told to him by Fraenkel himself.

[82]Igeret, Journal of the Israel National Academy of Sciences, no. 20, May 2001, page 25.

[83]Moshe Herr.

[84]Ein Harod is a kibbutz founded in 1921 by Russian Jews in the Jezreel valley.

[85]Igeret, May 2001, Journal of the Israel National Academy of Sciences, no. 20, page 25.

The End

On Tuesday, the 12th of October, 1965, Fraenkel gave a lecture on the mathematical calculations needed for the Jewish calendar to accommodate both lunar months and the solar year calendar. He also spoke about the historical development of the calendar. He was pleased with the well–attended lecture. Three days later, on the morning of Friday the 15th of October, he went swimming and exercising. At lunch he did not feel well and was taken to hospital. The ECG showed a minor attack, upon which Fraenkel said immediately, "I am ready to die." The physician responded that there was no need for that, and that he should just rest a while. Fraenkel responded that he had scheduled lectures at Bar Ilan University and the physician said he may need to be "somewhat late". As they wheeled him through the corridor from the emergency room, he had a massive heart attack and passed away instantly, the way he had always hoped to die.

Fraenkel wrote several wills. The first, written in 1938, stipulates that there will be no speeches during his funeral. He also forbade the use of foreign titles, Professor or Doctor, on his grave or a formal eulogy. He asked that his grave, like his father's, would have on it only the words "Of distinguished Jewish ancestry".

A longer will, dated the 22nd of November 1940, begins:

"Noting that no man knows his appointed time (especially at this fateful hour) therefore, being, thank God, of sound of mind and body, I do hereby declare to my family that this is my last will and testament." He continues to delegate his estate to his wife, and then adds a spiritual will:

> "You, my dear sons and daughters, I bless with a general blessing, that you will continue to walk in the path of *Tora va-avoda* (an orthodox Jewish life as well as a professional/vocational life); and that you will consider your private dealings secondary, whereas public matters, the nation's and the country's, and especially your family matters, shall be the main concern to which you should dedicate yourself. I wish you to have sons and daughters who would be absorbed in Jewish studies and in good deeds and employed in work that supports them and connects them with other people.
>
> Finally, know that worldly success is not the main objective. My grandfather, my father's father, was perhaps the most gifted of the last generations of my family, but he was reticent. My father, who was nearly as gifted and developed thanks to both his talents and his strong will (see his writings that I gathered in the book *Sigmund Fraenkel*) greatly outshone him and made a name and status for himself in Ashkenazi Jewry, both among economic leaders and in political echelons. I, who am inferior to both of them in talent and in strength of will and drive—though I have been blessed with both—have succeeded (at least until today) in my worldly success beyond both of them, especially thanks to the good luck that life in our country has given me and the rare fortune I have had to combine my talents for the benefit of the individual and the community. Deep is my gratitude to He who bestows wisdom for this success, but you should know, my sons, daughters and descendants: any worldly success has no value, and cannot provide contentment, if it is not grounded in respect for God and the offering of sacrifices for the common good: the family and the nation."

In 1950, on the day prior to *Yom Kippur* (20th of September), he wrote an appendage:

"An addition to my will of 1940 (with no derogation of the original). From the depth of my heart I feel gratitude to my parents, to my father, of blessed memory, and to my mother, may she enjoy a long life. Two outstanding advantages were bestowed upon me by my father: First, in a town, a community, and even in a family, where it was unusual, he passed down to me the basis of Jewish tradition, the Torah and the Talmud, through faithful teachers, our teacher Rabbi Hanoch Ehrentreu and our teacher Rabbi Eliezer Rotter, of blessed holy memory. This was based on the deep influence of his uncle Aviezri Zelig Auerbach, of blessed holy memory, the Rabbi of the Halberstadt community. He himself planted in me love for and devotion to the Creator (including prayers, *selihot* and *piyyutim*). In addition to the benefit to my soul, without this preparation I would not have been able to succeed in living a full Jewish life in our land. I would not have integrated linguistically or culturally in the vibrant life in the Jewish state. Second, my father was the one who swayed my heart towards a scientific profession, and helped me in some ways to find my path in this direction. I wonder whether I would have found the courage to choose the field of mathematics had it not been for my engagement with the Jewish calendar around the age of 15 (based on the foreword to the section on festivals in the *Mishnah* commentary "*Tiferet Israel*" which I found by chance.) This, together with the later support of my first steps by my uncle Alfred Loewi, of blessed memory, enabled me to perform successful independent research when I was still in school.

In my mother, I saw a model of an *Eshet Hayil,* a woman of valor,[86] a Jewish mother, who devotes her whole life to serving others. A great influence was also her mother, may she rest in peace (whose father's will is a live witness to a family tradition). And now here, God who creates a home for the lonely,[87] favored me with my wife, a daughter of sages of Israel from her father's and her mother's sides, and without whom I could not have succeeded as I have, neither internationally or in Israel. She too was a model of self-sacrifice to the benefit of others, whose longing for the Land of Israel already in her youth, together with her passion for Torah and *mitzvot*, prepared her specifically for life in our land. I thank you, my dear wife, for all that you have provided me, and I wish you a long and happy life among your descendants."

This will continues with a replication of the previous will and ends with:

"And I am, with a blessing from the bottom of my heart,

The word of the insignificant Avraham, son of the *Chaver* Aviezri HaLevi Fraenkel"

In 1963, he wrote:

"Addition to will

I add two comments to my will. These are not part of the will, but are rather informal matters, that I present to you at my seventy second birthday.

First, it seems to me that it would be commendable if my sons and daughters devote a significant portion of the inheritance, may be prior to the division of the property, to the high school education of our grandchildren. This includes, of course, also Miriam,[88] to the extent that her father has not already provided for this.

Second, as I now read a copy of my will, I am impressed at how sentimental I was at the time. This does not refer, of course, to the actual content of the will." ...

[86]Proverbs 31, 10.

[87]מושיב יחידים ביתה (Psalms 68, 7).

[88]Miriam is his granddaughter in the U.S. All other grandchildren were in Israel.

He then goes on to present the tolerant position described above, page 192.

In his eulogy of Fraenkel, Prof. Michael O. Rabin said, "When I think of him, three main attributes emerge: courage, humility, and love of people. When I say "courage", I do not mean physical courage—though he accomplished much in various sports, including mountaineering—which is surely a dangerous occupation—but I refer to a higher level of courage: the courage to figure out his stand and to stick to his point of view even when he was not in the majority. Like any energetic and successful person, Fraenkel knew how to collaborate with the powers and institutions that rule our world, so that he could promote the public and scientific interests he supported. But, when he was convinced, with rational and conscious conviction, with the integrity of a certain stand, he did not hesitate to take that stand even when it was against the opinion of the powerful or against the prevailing fashion. Fraenkel was absolutely unconcerned with what others thought of him, and in what he believed, he was fearless.

It seems to me that this courage was to a large extent based on humility. You could not find in him that constant preoccupation of people with their own self-importance. It is well known that Fraenkel was a world-renowned scientist, a researcher who contributed substantially to his field, a person who held high offices and had important roles. Yet despite all this, and may be because of this, there was no snobbery or vanity in him. The humility with which he spoke about himself and which he practiced in his work was a model for young people who came in contact with him.

The people who were close to him knew him as a person with a humane approach and an understanding heart. When he discussed any matter, he would think about the feelings of the people involved, and would weigh how this or that action would affect them. Many were assisted by him, and often his help was anonymous, so that the person receiving charity was unaware that it was Fraenkel who had assisted him. He was a loyal friend and wholly supportive. When he identified a genuine need, his loyalty was as solid and unshakeable as a rock. He would go through hell or high water to help a friend in need".[89]

[89]Rabin, M. On the scientific work of A.H. Fraenkel, in: B. Mazar, B., E. I. J. Poznanski, and M. Rabin. The late Professor Abraham HaLevi Fraenkel, presented on 1st December 65, published by Magnes Press, 1965.

Appendix

A.A. Fraenkel, *Recollections of a Jewish Mathematician in Germany*,
DOI 10.1007/978-3-319-30847-0

Fraenkel Family Tree

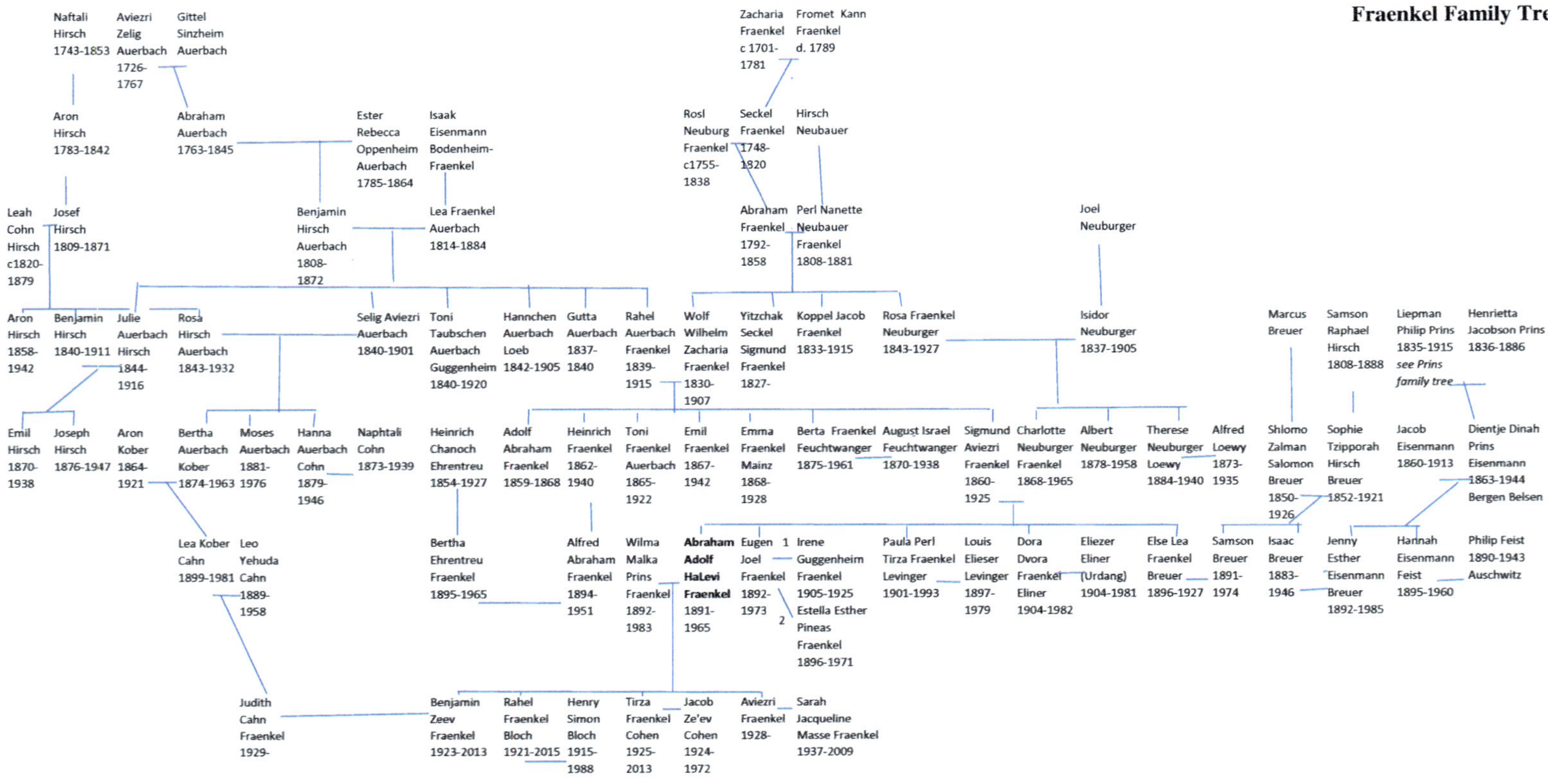

Note: The family tree is not complete and was designed merely to illustrate some of the family connections among the persons mentioned in the book.

c=circa d=died

Prins Family Tree

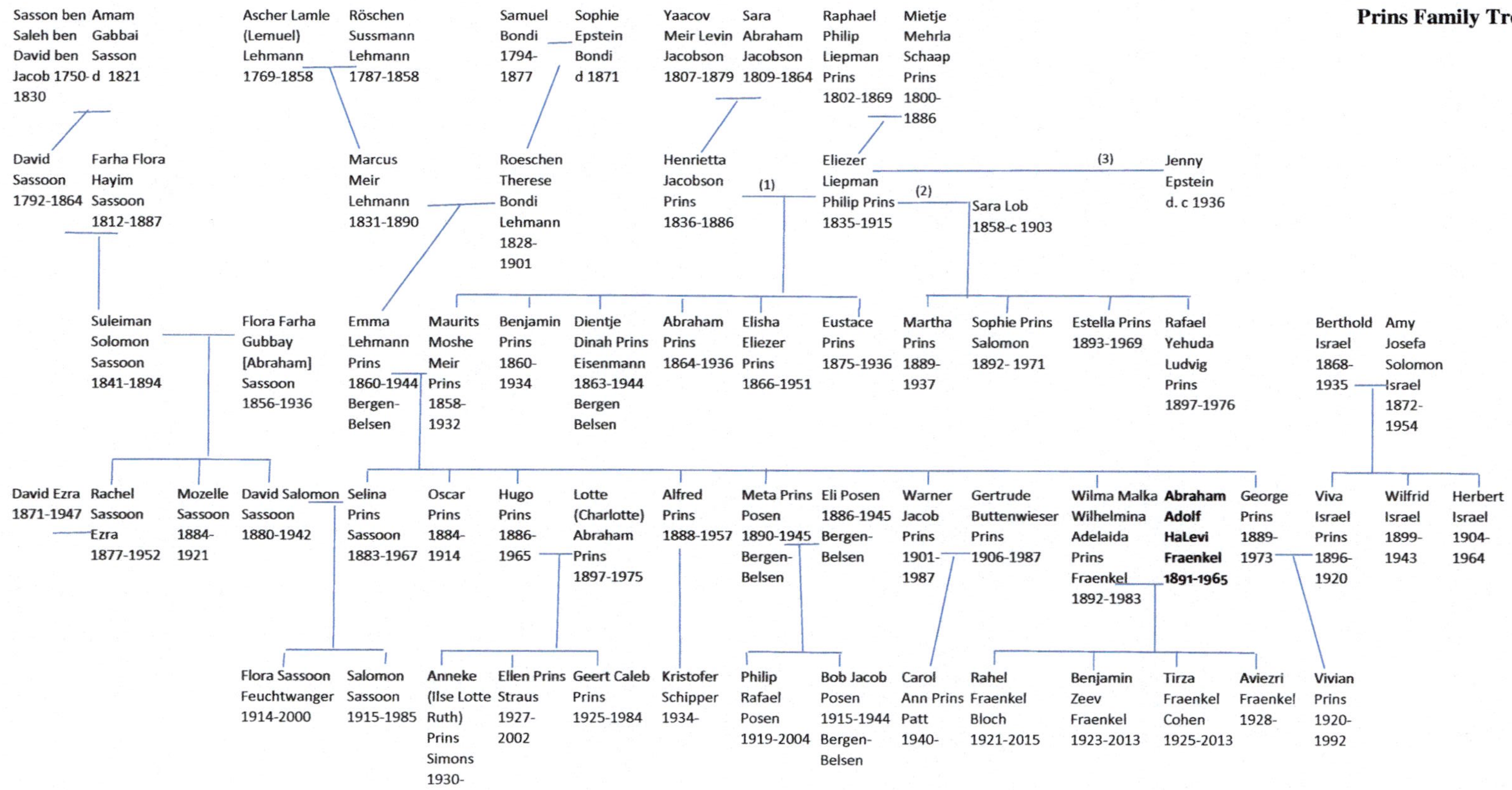

Note: The family tree is not complete and was designed merely to illustrate some of the family connections among the persons mentioned in the book.

c=circa d=died

Glossary

Glossary of Jewish, Hebrew, Yiddish Terms[1]

Acharonim **(=the last ones)** The leading Jewish exegetes and *poskim* (Jewish legal decisors) of the later (Jewish) Middle Ages, from the sixteenth century (specifically since the writing of the *Shulkhan Arukh*) to the present.

Adar Rishon The leap month, which is an extra (13th) month added to the Jewish calendar in certain years in order to make the Jewish year (a lunar calendar) maintain the seasons as in a solar calendar.

Adass Yisroel **(=congregation of Israel)** Refers to the Adass Yisroel Israelite Religious Congregation, the Orthodox secession congregation in Berlin, founded in June 1869. Its first rabbi was Dr. Azriel Hildesheimer. As a result of Edward Lasker's Secession Law of 1876, Adass Yisroel was converted from a "religious congregation" to a "synagogue congregation."

Aggadah **(=tale or lore)** The portion of the Talmud and *Midrash* that is not part of the Halacha and is thus not binding according to religious law. *Aggadoth* (plural): stories or sagas of the Talmudic and post-Talmudic period. *Aggadic* is the adjectival form.

Agudath Israel **(often referred to as *Agudah*)** World Agudath Israel, founded in May 1912 in Katowice as an organization of Orthodox Jews suspicious of secular Zionists, which first became politically active in 1919.

Almemar **(Arab. *al-minbar* = pulpit, Hebr. *bimah*)** Raised platform in a synagogue (usually at the center) from which the Torah is read and some of the prayers (according to Sephardic rites) are recited.

Amidah **("The Standing Prayer")** The central prayer of the Jewish liturgy.

Ashkenazim In a narrow sense German and Dutch Jews, but in its wider sense commonly used today: Jews of Europe (except for southern Europe) and North

[1]Spelling generally follows the Sephardic pronunciation, except in cases such as *Adass Yisroel*, in which for historical reasons the Ashkenazi pronunciation was deemed more appropriate.

A.A. Fraenkel, *Recollections of a Jewish Mathematician in Germany*,
DOI 10.1007/978-3-319-30847-0

America who are related by virtue of their rites and to some extent the use of the Yiddish language (cf. *Sephardim*).

Av The 11th month in the Jewish calendar, corresponding to July–August (depending on its position in the leap year cycle). *Tisha b'Av* (lit: Ninth of Av): a day of fasting and mourning in commemoration of the destruction of the First and Second Temples in Jerusalem.

Bar Mitzvah **(=son of the commandment, subject to the commandments)** A boy after reaching the age of 13, when he assumes the duties and responsibilities of an adult according to religious law. The ceremony and celebration of a boy having reached this age is also referred to as a *bar mitzvah*.

Baal koreh **(*leiner*)** The person who reads the Torah in synagogue as part of the service.

Chaver **(=comrade, friend)** Title that especially in Germany was conferred in recognition of a person's Talmudic study and knowledge or service to the community. See: http://www.daat.ac.il/encyclopedia/value.asp?id1=3145

Chazan Jewish prayer leader or cantor.

Chumash Pentateuch, or the Five Books of Moses.

Derashah **(pl. *derashot*) (=discourse, homily)** Scholarly speech or sermon, usually preached by a rabbi or a bar mitzvah boy.

Derekh ha-Limmud Method of Talmud (etc.) study, as applied in the respective country or school (usually used in masculine form, but feminine is more correct).

Emek Valley, wide plain.

Erev **(=evening, eve)** Term referring to the day preceding a holiday or day of fasting.

Gaon Exceptional scholar (Talmudist).

Gemara **(from *g'mor* [Aramaic *gamar*] = to learn, study)** The more recent portion of the Talmud (starting in the third century CE), which comprises analysis of and commentary on the Mishnah; usually refers to the Babylonian Gemara.

Halacha The standard according to religious law; the subject matter of religious law, particularly in the Talmud. *Halachic* is the adjectival form.

Haredim Orthodox Jews characterized by strict observance of Jewish law and custom and often by rejection of secular culture.

Hasid **(pl. *hasidim*) (=pious)** Follower of the doctrine and movement of traditional Orthodox Judaism founded by Rabbi Israel Baal Shem Tov (1699–1760), which had spread among wide circles of Ashkenazi Jewry (especially in Poland and Galicia) starting in the mid-eighteenth century.

Hoshana Rabbah The seventh and final day of *Sukkot*.

Kaddish Aramaic prayer recited during religious services and by those in mourning.

Kasher (verb) To render kosher; to boil, heat to glowing, etc. in order to make kitchen and other appliances fit for ritual use.

Kashrut (noun) Observance of the ritual dietary laws; the state of being kosher.

Kibbutz **(pl. *kibbutzim*)** Collective settlement.

Kol Nidre The prayer introducing the evening service of *Yom Kippur* according to Ashkenazi tradition; often used to refer to the full evening service which starts *Yom Kippur*.

***Kosher* (adj) (kasher = fit for use)** Corresponding to the ritual dietary laws.

***Leining* (from Latin *legere*)** Reading from the Torah (that is, the weekly parashah, or portion) in the synagogue, using traditional *trop* (Yiddish "cantillation" or "trope").

***Luach* (=tablet)** The Jewish calendar.

***Mahzor* (=cycle)** Prayer book containing the prayers and *piyyutim* (see below) for one or more holidays (cf. *siddur*).

Matzoh Unleavened bread, as prescribed for Passover (*Pesach*).

Megillat Esther The biblical Book of Esther, often in the form of a hand-written scroll (Hebr. *megillah*).

***Midrash* (pl. *midrashim;* adj. *midrashic*)** An ancient manner of expounding the Bible which includes interpretations and stories that add to and fill in gaps in the original text.

Mincha Afternoon prayer service.

Minhag Liturgical ritual or religious custom.

***Mishnah* (pl. *mishnayot*)** The older portion of the Talmud, written in the last two centuries BCE and the first two centuries CE.

***Mitzva* (pl. *mitzvot*)** A precept or commandment of which Judaism has 613.

Mizrachi Party (Organization) of Orthodox Zionists.

Musaf The "additional" prayer recited on the Sabbath, the new month (*Rosh Chodesh*), and on holidays.

***Ner Tamid* (=eternal light)** The permanently lit sanctuary lamp that hangs in all synagogues in front of the Torah ark.

***Neturei Karta* (=Guardians of the City)** The ultra-Orthodox sect, mainly in Jerusalem, New York, and London, which radically opposes Zionism and refuses to recognize the existence or authority of the State of Israel.

***Ohel Jakob* (=Jacob's tent)** Name of the Orthodox congregation in Munich within the framework of the entire unified Jewish congregation.

***Omer* (counting of the Omer)** Omer is a measure of volume, but also signifies the sheaves of grain, for example, barley; the counting of the 49 days from the second day of Passover (marking the ripening/harvesting and offering of the barley) until *Shavuot* (Feast of Weeks).

Oriental Jews (Descendants of) Jews of North Africa and the Middle East.

***Parashah* (=portion)** Weekly portion of the Pentateuch that is read from the Torah in the synagogue on Shabbat, so that the Pentateuch is completed on Simchat Torah and started anew from the beginning the following Shabbat.

***Pesach* (=passing over)** Passover, the seven-day festival commemorating the liberation and exodus of the Jews from Egyptian slavery, beginning on the 15th of Nissan, usually starting in the week preceding the Christian Easter celebration, cf. *Seder*.

***Piyyut* (pl. *piyyutim*, from the Greek *poietes*)** Liturgical poems that extend the regular prayers on holidays, written from approximately the eighth century until the late Middle Ages.

***Purim* (=lots)** The joyous festival commemorating the survival of the Jews of Persia (see *Megillat Esther*), celebrated one month before Passover.

***Rashi* (acronym for Rabbi Shlomo Yitzchaki, died 1105)** The traditional exegete of the Bible, especially the Pentateuch, and the classic commentator of the Talmud.

Rebbe A Yiddish word derived from the word rabbi, which means "master, teacher, or mentor." It refers to teachers of Torah.

***Rosh Hashanah* (=beginning of the year)** The first and second day of the month of *Tishrei*, in September or early October; a biblical holiday, later celebrated as the Jewish New Year.

***Seder* (=order)** The first evening (in the Diaspora the first two evenings) of the Passover celebration, characterized by a lengthy festival liturgy and feast, which commemorates the exodus of the Jews from Egypt, which took place on that day, cf. *Pesach*.

***Seder Mo'ed* (=festival order)** The second of six sections of the Mishnah or Talmud, which deals primarily with Shabbat and the holidays.

***Selichot* (=forgiveness)** Prayers and poems of repentance for the Day of Atonement (*Yom Kippur*) and the preceding weeks, as well as for other days of fasting.

Sephardim In a narrow sense descendants of the Jews expelled from Spain and Portugal in the late fifteenth century, but in its wider sense Oriental (and southern European) Jews, except for Yemenite Jews (cf. *Ashkenazim*, Oriental).

***Shabbat* (Sabbath)** Saturday. The seventh day of the week, celebrated as a day of rest.

Shabbat HaChodesh The Sabbath between 14 and 19 days before Passover, characterized by an additional Torah reading (Exodus 12:1–20) that deals with the festival.

Shacharit Morning prayer service.

Shemini Atzeret The eighth day of the *Sukkot* festival, and also an individual holiday in its own right.

Shivah The traditional seven-day period of mourning following the funeral of a close relative.

Shochet A professional ritual (kosher) slaughterer of animals.

***Shul* (Yiddish; from German: *Schule*, school)** Synagogue; *Bachurim Shul*: Synagogue of the students.

***Siddur* (= order)** Prayer book for every day and Saturdays (cf. *mahzor*).

***Simchat Torah* (=rejoicing in the Torah)** The holiday following Shemini Atzeret. In Israel, Shemini Atzeret and Simchat Torah are celebrated together.

Sukkot The Feast of Booths (Tabernacles), starting on the 15th of Tishrei. *Sukkah* (sing.) = booth.

Talmud The major compendium of Oral Law (as opposed to the written law contained in the Bible), written partly in the Land of Israel and partly in Babylon from the second century BCE to the fifth century CE (cf. *Mishnah* and *Gemara*).

Tefillin Phylacteries; leather boxes containing scriptural verses tied onto the left arm and forehead by Jewish men during weekday morning prayers.

"Three weeks" The three weeks from the day of fasting on the 17th of Tammuz, marking the day the Babylonian army stormed Jerusalem's city walls, until the particularly solemn day of fasting on Tisha b'Av (destruction of the First and Second Temples by the Babylonians and the Romans); traditional weeks of mourning.

Tiferet Yisrael Commentary on the Mishnah written by Israel ben Gedalia Lipschitz (1782–1860).

***Torah* (=instruction)** In a narrow sense, the Pentateuch (Five Books of Moses); in a broader sense, Jewish religious teachings in general.

***Tosafot* (=additions, i.e. to Rashi; sing. *tosafa*)** Commentaries to the Talmud written by scholars in France and Germany in the twelfth and thirteenth centuries.

Trop Yiddish word for the musical cantillation marks for the ritual chanting of the Torah in the synagogue, often called "trope" in English; cf. *leining*.

Yekke Derogatory or humorous term for Jews from Germany.

***Yeshiva* (=sitting)** High (or middle) school for Talmudic study; the term is also used for the institutions in which adults study Jewish texts as their main vocation.

Yishuv The body of Jewish residents in Palestine/Land of Israel prior to the establishment of the State of Israel.

Yom Kippur The Jewish Day of Atonement, on the 10th of Tishrei.

***Yom Tov* (="good day")** Holiday or festival.

Glossary of German Educational Terms: School Level

Abitur The university entrance qualification examination after completion of secondary school.

Höhere Töchterschule Girls' secondary school.

Gymnasium Academic (university-track) secondary school, usually with a specific focus, either in the humanities or natural sciences.

Mittelschule Secondary schools in Bavaria.

Oberrealschule Natural science-oriented secondary school.

Realgymnasium Secondary school that focused on mathematics and modern languages rather than Greek and Latin.

Simultaneous school Public school in Germany that was not associated with a specific religious denomination, attended by both Catholic and Protestant students.

Volksschule Regular primary school.

Vorschule Preparatory school affiliated with the secondary school.

Glossary of German Educational Terms: University Level

(descriptions are only approximate equivalents to English-language university teaching positions)

Extraordinarius Non-tenured associate professor.

Habilitation Postdoctoral professorial qualification, granting the person the *venia legendi* (permission to lecture/teach) as a full professor at a university.

Ordinarius Full professor or department chair.

Privatdozent Unsalaried, adjunct professor; has completed the postdoctoral professorial qualification and must teach each year to retain the status; does not (yet) have a paid position as a full professor.

Professor ad personam Professor with a personal chair established specifically for that person.

Bibliography[2,3]

Fraenkel AA (1908) Bestimmung des Datums des jüdischen Osterfestes für die Zeitrechnung der Mohammedaner. Zeitschrift für Mathematik und Naturwissenschaft Unterricht 39(6):598–606

Fraenkel A (1909) Eine Formel zur Verwandlung jüdischer Daten in mohammedanische. Monatsschrift für Geschichte und Wissenschaft des Judentums 53:70–79, Also printed in Jeder Bundesbruder bezieht seine Bücher vom hebräischen Verlag Menorah. Berlin, no. 6, pp. 736–743. Also printed in Munich 1910 by Dreck von Adolf Alkalay & Sohn, Pressburg, pp. 1–10. http://sammlungen.ub.uni-frankfurt.de/cm/periodical/pageview/2889720 (German)

Fraenkel AA (1910) Die Berechnung des Osterfestes. Journal für die reine und angewandte Mathematik 138:133–146 (German)

Fraenkel AA (1911) Lösung einer Aufgabe von P. Wolfskehl. Archiv der Math u Phys 17(3):92–94

Fraenkel AA (1911) Le calcul de la date de Pâques. Bologna, N. Zanichelli. 7p. Estratto da 'Scientia' rivista di scienza, vol IX, anno V

Fraenkel AA (1912) Axiomatische Begrundung von Hensels p-adischen Zahlen. Journal für die reine und angewandte Mathematik 141:43–76

Fraenkel AA (1914) Über die Teiler der Null und die Zerlegung von Ringen: Inaugural-Dissertation zur Erlangung der Doktorwurde der hohen philosophischen Fakultät der Universität Marburg. Marburg (Georg Reimer, Berlin) 42, [41] p. J f Math 145(1915):139–176

Fraenkel AA (1915) Über die Osterrechnung. Quelle & Meyer, Leipzig, pp 264–272

Fraenkel AA (1916) Jüdische Eindrücke im Felde (Jewish impressions from the field). Frankfurter Israelitisches Familienblatt (FIFB) 14(12):1–2. http://sammlungen.ub.uni-frankfurt.de/cm/periodical/pageview/2704603

Fraenkel AA (1914–1919) Gedanken über künftige Entwicklung und Aufgaben der Agudas Jisroel (Thoughts on future developments and tasks of the Agudas Yisroel) in an unknown source in the Agudist and/or Zionist press in Germany

Fraenkel AA (1916) Thoratreues Judentum und zionistische Organisation (Torah-faithful Judaism and Zionist organization). Der Israelit 44:1f

[2]This bibliography was not included in the original autobiography. Rather, it was compiled from multiple sources. Although references were checked whenever possible, the bibliography should not be considered comprehensive, as it was impossible to locate all articles written by Fraenkel.

[3]Writings for which a copy exists in Fraenkel's archive at the National Library of Israel ARC 4# 1621 AC2060 are marked with a hash sign (#). Those which are in Fraenkel's archive at the National Library of Israel ARC 4# 1621 AC-2054 are marked with two hash signs (##).

A.A. Fraenkel, *Recollections of a Jewish Mathematician in Germany*,
DOI 10.1007/978-3-319-30847-0

Fraenkel AA (1916) Über gewisse Teilbereiche und Erweiterungen von Ringen. B.G. Teubner Verlag, Leipzig, 64pp. Also in *Monatshefte für Mathematik*, 28, no. 1, Dec 1917, p.A29(1)

Fraenkel AA (1918) Praktisches zur Universitätsgründung in Jerusalem. Der Jude 3:404–414. http://sammlungen.ub.uni-frankfurt.de/cm/periodical/titleinfo/3103705

Fraenkel AA (1918) Kursberechnung für Flugzeuge. Grosses Hauptquartier, 15pp

Fraenkel AA (1918) Mathematik und Apologie. Jeschurun 5:112–126

Fraenkel AA (1919) Einleitung in die Mengenlehre: eine gemeinverständliche Einführung in das Reich der unendlichen Grössen. Springer, Berlin. Book. 156pp

Fraenkel A (1919) Von Nah und Fern [From near and far] (German). Die Jüdische Presse, Berlin, 11 July 1919

Fraenkel AA (1919) Die sexuelle Frage im künftigen Palästina (The question of sex in the future Palestine). Jüdische Rundschau 24(19):142–143. Also published in *Jüdische Presse*. http://sammlungen.ub.uni-frankfurt.de/cm/periodical/pageview/2669686

Fraenkel AA (1920) Zahlbegriff und Algebra bei Gauss. 1. Teil. Mit einem Anhang von A. Ostrowski. Materialien fur eine wiss. Biographie von Gauss, No. VIII. Also in Nachrichten von Königlichen Gesellschaft der Wissenschaften zu Göttingen, Math-Phys. Kl. Göttingen, 59pp (Appeared also in B.G. Teubner, Leipzig)

Fraenkel A (1920) Das Erziehungswesen in ארץ ישראל u. der Misrachi [The educational system in the land of Israel and the Mizrachi movement]. Die Jüdische Presse, Nr 9, Berlin, 27 Feb 1920 (German) #

Fraenkel A (1920) Von Amsterdam nach London [From Amsterdam to London]. Die Jüdische Presse, Nr 46, Berlin, 12 Nov 1920 (German) #

Fraenkel AA (1921) Über einfache Erweiterungen zerlegbarer Ringe. Journal für die reine und angewandte Mathematik 151:121–167

Fraenkel AA (1921) Über die Zermelosche Begründung der Mengenlehre. Jahresbericht der Deutschen Mathematikervereinigung 30:97–98

Fraenkel A (1921) Zur Politik des Misrachi. Die Jüdische Presse, Berlin, 29 July 1921 (German) #

Fraenkel A (1922) Axiomatische Begründung der transfiniten Kardinalzahlen I. Mathematische Zeitschrift 13:153–188

Fraenkel AA (1922) Den Begriff "definit" und die Unabhängigkeit des Auswahlaxioms. Sitzungsb d Preuβ Akad d Wiss, Physik-math. Klasse, pp 253–257. (Abstract in *Jahresb. D. M. V.* 31 (1922), 101–102 ital.) see also http://biodiversitylibrary.org/page/29207376#page/276/mode/1up

Fraenkel AA (1922) The notion of "definite" and the independence of the axiom of choice. Translated by Beverly Woodward. Harvard University Press, Cambridge, MA, pp 284–289

Fraenkel AA (1922) Der Zusammenhang zwischen dem ersten und dem dritten Gausschen Beweis des Fundamentalsatzes der Algebra. Jahresb D M V 31:234–238

Fraenkel A (1922) Zu den Grundlagen der Cantor-Zermeloschen Mengenlehre (On the foundations of the Cantor-Zermelo set theory). Mathematische Annalen 86:230–237 (Abstract in *Jahresber. D.M.V.* vol 30, p. 97–98 ital., 1921)

Fraenkel AA (1923) Einleitung in die Mengenlehre: eine elementare Enführung in das Reich des Unendlichgrossen, 2nd ed. Springer, Berlin, Book. 251pp (Die Grundlehren der Math Wissenschaften, vol IX)

Fraenkel AA (1924) Die neueren Ideen zur Grundlegung der Analysis und Mengenlehre. Jahresbericht der Deutschen Mathematiker-Vereinigung 33:97–103 (Cf. *Scientia*, vol. 38, 141–152, 209–218, 1925. Also in French)

Fraenkel AA (1924) The Jewish University in Jerusalem (From the viewpoint of orthodoxy). Jewish Forum VII(1):27–31

Fraenkel AA (1924) The Jewish University in Jerusalem (From the viewpoint of orthodoxy). Jewish Forum VII(5):299–302

Fraenkel A (1924) Die jüdische Universität in Jerusalem. Jüdische Wochenschrift, Nr 16/17, Vienna, 18 Apr 1924 (German) #

Fraenkel AA (1924) האכסיומות של תורת הקבוצים. בתוך: כתבי האוניברסיטה ובית הספרים בירושלם. עורך א. אינשטיין. כרך א'. התרפ"ד. תרגם בנימין אמירה. The Axioms of set theory. In: Einstein A (ed) The texts of the University and the Library in Jerusalem. Mathematics and Physics vol 1. Scripta Universitatis Atique Bibliothecae Hierosolymitanarum : Mathematica et Physica vol 1. Parallel Hebrew title: Kitve ha-Universitah u-Vet ha-Sefarim bi-Yerushalayim : Matimatikah u-Fisikah. Fraenkel articles in both Hebrew and German. Hebrew is on pages: 1–5, German is on pages 1–8. In both languages it is listed as article VI. German article is titled: Die Axiome der Mengenlehre; Hebrew title is: ha-Aksiyomot shel torat ha-kabutsim

Fraenkel AA (1924?) Vom Misrachi in Wiesbaden. Jüdisches Wochenblatt 2:15. (German) # [Question mark after the year in references was copied from Fraenkel's note on the article.]

Fraenkel AA (1924) Das neunte Gebot (The ninth commandment). Jüdisches Wochenblatt 13:115 (German) #

Fraenkel AA (1925) Die hebräische Universität, der 'Israelit' und כבוד התורה, (The Hebrew University, the "Israelit" and the honor of the Torah). Jüdisches Wochenblatt 6:60–61. http://archive.org/stream/jdischeswochenblreel01#page/n246/mode/1up/search/fraenkel

Fraenkel AA (1925) Über die gegenwärtige Grundlagenkrise der Mathematik. Koch, 10p

Fraenkel AA (1925) Untersuchungen über die Grundlagen der Mengenlehre. Mathematische Zeitschrift 22:250–273

Fraenkel A (1925) Leben, Natur, Religion. Jeschurun 12:337–348 (German). Also as a separate monograph, pp 1–12. #

Fraenkel A (1925) Bemerkung zum Begriff der geordneten Mengen. Fund Math 7:308–310 (Enlarged in *Sitzungsber. der Berliner Math. Gesellschaft.* 24, 29–37)

Fraenkel A (1926) Axiomatische Theorie der geordneten Mengen. Journal für die reine und angewandte Mathematik 155:129–158

Fraenkel A, Vieler H (1926) Untersuchungen über Unabhängigkeit und Tragweite der Axiome der Mengenlehre in der Axiomatik Zermelos und Fraenkels. (Göttingen, Dieterich). 41p 8°. Diss, Marburg

Fraenkel A (1926?) Briefe an die Redaktion [Letter to the editor.] An unknown Jewish newspaper. N63, p 22 (German) #

Fraenkel A (1926) Von jüdischen Reisen, Einige praktische Erfahrungen einer Palästinareise. Jüdisches Wochenblatt, Nr 17 May 1926, p 157 (German) #

Fraenkel AA (1927) Zehn Vorlesungen über die Grundlegung der Mengenlehre. Wissensch Und Hypothese, vol 31. B.G. Teubner Verlag, Leipzig, Book 182pp

Fraenkel A (1927) Die Gleichheitsbeziehung in der Mengenlehre. Journal fur die reine und angewandte Mathematik 157:79–81

Fraenkel A (1927) Rabbiner Dr. H. Eherentreu. Jüdischer Wochenblatt 1:5 (German) #

Fraenkel A (1927) Letter to the editor from Marburg, 30th November. Section: Ferienrabatt für alle! In: Beilage zur Vossischen Zeitung (German) #

Fraenkel A (1927) Warum nicht mehr Schlafwagen 3. Klasse? [Why are there no more third class sleeping cars (on the train). 31 Mar 1927 (German) #

Fraenkel A (1928) Zusatz zu vorstehendem Aufsatz Herrn v. Neumanns. Mathematische Annalen 99:392–393

Fraenkel AA (1928) Einleitung in die Mengenlehre, 3rd ed. Springer, Berlin, 424pp. Book

Fraenkel A (1928) Über die Ordnungsfähigkeit beliebiger Mengen. Sitzungsber d Preuß Ak d Wiss, Phys-math, Berlin, pp 90–91

Fraenkel A (1928) Gelöste und ungelöste Probleme im Umkreis des Auswahlprinzips. Bologna, Zanichelli. (Estratto dagli Atti del Congresso intemazionale dei matematici, Bologna, vol VI:255–259, 1930)

Fraenkel A (1929) Mathematics and philosophy. Opening Address 18th November 1929, HUA, EIM files, Stencil, 4p [Hebrew]

Fraenkel A (1929) Misrachisten in der Agudah? [Mizrachi Members in the Agudah?]. Der Israelit Blätter 12:3 (German) #

Fraenkel A (1929) Der Sinn des Delegiertentages des Misrachi. Jüdisches Wochenblatt 7/8:472. Berlin, 22 Feb 1929 (German) #

Fraenkel A (1929) Sprechsaal. Zweierlei Misrachismus [Two types of Mizrachi movement]. Der Israelit. 8 Aug 1929 (German) #

Fraenkel AA (1930) Georg Cantor. Jahresbericht der Deutschen Mathematiker-Vereinigung 39:189–266. Also appeared separately as *Georg Cantor* Leipzig: B. G. Teubner and is abridged in Cantor's *Gesammelte Abhandlungen* (Cf. Scientia, vol 48, 361–370, 1930; also appeared in French)

Fraenkel AA (ed) (1930) Sigmund Fraenkel: Aufsätze und Reden; ein Spiegelbild deutsch-jüdischer Geschichte aus dem Beginn des zwanzigsten Jahrhunderts. B. Heller, Munich, 343pp (German)

Fraenkel AA (1930) Die heutigen Gegensätze in der Grundlegung der Mathematik. In: Erkenntnis, vol 1, pp 286–302 (Cf. Ztschr. F. math. Und naturwiss. Unterricht, vol. 61, p. 35, 1930; Blätter f. Deutsche Philos., vol. 4, 279–297, 1930; Scr. Math., vol. 1, 222–227, 1934)

Fraenkel A (1930) Das Werden der hebräischen Universität in Jerusalem [The formation and development of the Hebrew University in Jerusalem]. Unterhaltungsblatt Kölnische Zeitung, 22 July 1930 (German) #

Fraenkel A (1930) באוניברסיטה העברית [At the Hebrew University]. הארץ כ"א מרחשון גליון issue 3428. *Ha'aretz* 12 Nov 1930 (Hebrew) #

Fraenkel AH (1930–1931) (5691) "אמונות ודעות לאור מדעי הטבע" [Beliefs and opinions in light of the sciences]. Part 1 in *HaHed* ההד VI(8):16–19, part 2 in ההד *HaHed* VI(9). Reprinted together as a monograph by *HaHed* ההד in 1930 15 תרצ"א pp. דפוס הספר ירושלים HaSefer Printing Press Jerusalem (Hebrew) #. Reprinted in 1987–1988. (Hebrew). Translated into English by Mark Zelcer in *Hakirah* the Flatbush Journal of Jewish Law and Thought, 2011, vol 12:220–229

Fraenkel A (1931) Introduction to W.M. Feldman. Rabbinical mathematics and astronomy, 3rd corrected edition. Hermon Press, New York, NY. Also 1965 and 1978

Fraenkel A (1931) Beschränkung fördert Durchführung. [Restriction promotes implementation]. Die Brücke. Des Berliner Tageblatts, Nr 51, 27 Dec 1931 (German) ##

Fraenkel A (1931) Austauschstudium an der hebräischen Universität [Exchange studies at the Hebrew University]. Jüdische Rundschau, Nr 88, 13.XI.1931 (German) #

Fraenkel A (1931) Ein Vorschlag [A suggestion]. Israelitisches Familienblatt, vol 33. Hamburg, 10 Dec 1931 (German) #

Fraenkel AA (1932) Axiomatische Theorie der Wohlordnung. Journal für die reine und angewandte Mathematik 167:1–11. http://gdz.sub.uni-goettingen.de/dms/load/img/?PPN=PPN243919689_0167&DMDID=DMDLOG_0006&LOGID=LOG_0006&PHYSID=PHYS_0008

Fraenkel AA (1932) Das Leben Georg Cantors. In: Ernst Zermelo (Ed). Georg Cantors Gesammelte Abhandlungen, Berlin, pp 452–483

Fraenkel A (1932) Bei den palästinensischen Studenten in Jerusalem [The Palestinian students in Jerusalem]. Neues Wiener Journal. Vienna, 10 Mar 1932 #

Fraenkel A (1932) Die Orthodoxie und der Aufbau Palästinas. Jüdische Presszentrale 712:13 (German) #

Barth A, Wolfsberg O, Fraenkel AH, Struck H, Shapira Y (1933) .גלוי דעת בענייני הקרנות [Opinion regarding the funds]. דבר Davar, p 5. 11 Dec 1933

Fraenkel AA (1933) On modern problems in the foundations of mathematics. Scripta Math I:222–227

Fraenkel A (1934) An den Gestaden des toten Meeres [On the shores of the Dead Sea]. Jüdische Rundschau, Nr 96, 30.XI.1934 (German)

Fraenkel A (1934) Palästina Probleme. Jüdische Presszentrale 803:9 (German) #

Fraenkel A (1935) Sur l'axiome du choix. Fondation L'Enseignement Mathématique. 34me annèe, pp 32–51 (Cf. Kongr. Zurich 1932, vol. II, 341 f., 1933; Revue Philos. De Louvain, vol. 50, 429–459. 1952. http://retro.seals.ch/digbib/view?pid=ensmat-001:1935:34::193)

Fraenkel AA (1935) Sur la notion d'existence dans les mathematiques : sur l'axiome du choix. Gauthier-Villars, Paris

Fraenkel AA (1935) Zum Diagonalverfahren Cantors. Fund Math 25:45–50

Fraenkel AA (1935) Concerning the method of number pairs. Philos Sci 2(1):104–108

Fraenkel A (1935) Sur la Notion d'Existence dans les Mathematiques'. L'Enseignement Mathematique 34:18–32

Fraenkel A (1935) "על עבודתו של הרמב"ם בשדה 'סוד העבור' והתכונה" [on the Rambam's work on the topics of the secret of leap months and in astronomy]. In: "רבנו משה בן מיימון למלאות שמונה מאות שנה להולדתו" [Our Rabbi Moshe ben Maimon. 800 years from his birth]. *Ha'aretz* (4788):111–114. Tel Aviv (Hebrew)

Fraenkel A (1936) פרחה (פלורה) ששון ז"ל [Farcha (Flora) Sassoon]. *הארץ* Ha'aretz, 24 Jan 1936 (Hebrew) #

Fraenkel A (1936) .הרמן שטרוק [Hermann Struck]. *הארץ* Ha'aretz, 6 Mar 1936 (Hebrew) #

Fraenkel AA (1937) Über eine abgeschwächte Fassung des Auswahlaxioms. J Symb Logic 2(I):1–25 (Abstract in *C.R. de l'acad. Des Sc.* Paris, vol. 192, (1931) p. 1072; Jahresb. D.M.V., vol 41, p. 38, 1932?)

Fraenkel A (1937) Rudolf Otto Eulogy. Ha'aretz, 7 Mar 1937

Fraenkel A (1937) Science at the Hebrew University "An aid to national development". The Jewish Chronicle, 29 Oct 1937, p 29 (English) #

Fraenkel A (1937) Discontinu et continu (IXme Congrès Intern. de Philosophie). Actualités Scientif et Industr 535:193–200 (Cf. Ztschr. F. Freie Deutsche Forschung, vol. 2, 1–14, 1939; Scr. Math., vol. 13, 17–36, 1947)

Fraenkel AH (1937) עמדת היהודים "לריפורמה על הלוח" The Jews' position regarding the "Calendar Reform". *הארץ* Ha'aretz, 11 June 1937 (Hebrew) #

Fraenkel AH (1937) Letter to the editor of *לשוננו לעם Leshonenu Le'Am* [Our language for the people]. 8 Oct 1937 (Hebrew) #

Fraenkel AH (1937) Letter to the editor [concerning the Mizrachi and the Poel HaMizrachi movements]. *נתיבה* Netiva. 19 ט"ו בכסלו תשצ"ח Nov 1937 (Hebrew) #

Fraenkel AA (1938) Alfred Loewy (1873–1935). Scripta Math V(1):17–22

Fraenkel AH (1938) יחזקאל הלוי (אדמונד) לנדאו ז"ל [Eulogy] Yehezqel Ha-Levi (Edmund) Landau of Blessed Memory. *הארץ* Ha'aretz, 4 Mar 1938 (Hebrew) #

Fraenkel AH (1938) .לימודי הטבע באוניברסיטה העברית ותורת ישראל [The study of science at the Hebrew University and Jewish religious teachings]. *הצופה* Ha'Tzofe, 10 Apr 1938 (Hebrew) #

Fraenkel AA (1939) Address of the rector of the Hebrew University, A.H. Fraenkel: graduation ceremony 22 Nissan 5699. Azriel Press, Jerusalem, 11p

Fraenkel AA, Bar-Hillel J (1939) Le Problème des antinomies et ses développements récents. Revue de Métaphysique et de Morale 46:225–242

Fraenkel AA (1939) Natural numbers as cardinals. Scripta Math VI(2)

Fraenkel A (1939) Review: Adolf Lindenbaum, Andrzej Mostowski, Über die Unabhängigkeit des Auswahlaxioms und einiger seiner Folgerungen. J Symb Logic 4(1):30–31

Fraenkel AH (1939) המקרה הוא? *ההד*. עמודים י-יא' תשרי-מרחשון ת"ש [Is it by chance?] HaHed 15(1–2):10–11 (Hebrew)

Fraenkel AH (1939) בענין "ספר האשכול" [Concerning "Sefer HaEshkol"] *הארץ* Ha'aretz, 31 Jan 1939 (Hebrew) #

Fraenkel AH (1939) .טקס הסיום באוניברסיטה העברית. נאום הריקטור פרופסור. אברהם הלוי פרנקל [Graduation ceremony at the Hebrew University. The Rector's address].*הצופה*. Ha'Tzofe, 13 Apr 1939 (Hebrew) #

Fraenkel AH (1939) לתקנתה של התרבות העברית [Enhancing Hebrew culture]. *הארץ* Ha'aretz, 28 May 1939 (Hebrew) #

Fraenkel AH (1939) .הרב אברהם אליהו קאפלאן חמש עשרה שנה לפטירתו. תולדותיו [Rabbi Abraham Eliyahu Kaplan, 15 years after his passing]. *הצופה* Ha'Tzofe, 5 י"ז באייר, תרצ"ט May 1939 (Hebrew) #

Fraenkel AH (1939) מהפכות במדע [Revolutions in science]. *הצופה* Ha'Tzofe, 14 Apr 1939 (Hebrew) #

Fraenkel AA (1940) Review: Josef Schachter, H. Bergmann, Brief introduction to logistic. J Symb Logic 5(1):43

Fraenkel AA (1940) Natural numbers as ordinals. Scripta Math VII(1–4):9–20

Fraenkel AH (1940) אישיותו של אלפרד ברגר ז"ל [The late Alfred Berger]. *הארץ*. ר"ח שבט ת"ש Ha'aretz, 11 Jan 1940, p 3 (Hebrew) #

Fraenkel AH (1940) Science at Hebrew University. The Australian Jewish News. Melbourne, 31 May 1940 (English) #

Fraenkel AH (1940) נאום רקטור האוניברסיטה, פרופ' א.ה. פרנקל בטכס הסיום. [The address of the University Rector, Prof. A.H. Fraenkel in the graduation ceremony]. Ha'Tzofe, 5 May 1940 (Hebrew)

Fraenkel AH (1940) "מכתב למערכת". Letter to the editor [concerning reciting of Selichot]. *הצופה* Ha'Tzofe, 7 Oct 1940, p 2 (Hebrew) #

Fraenkel AH (1941) "מכתב למערכת". [Letter to the editor concerning the date for Yom Kippur in the far east]. *הצופה* Ha'Tzofe, 12 Sept, p 8 (Hebrew)

Fraenkel AH (1941) הוכחות של אי אפשרות במדע המדויק [Proving impossibility in exact sciences]. *טכניקה ומדע. ירחון למדע שימושי ולהשכלה.* Tech Sci. Appl Pure Sci Month 5(2):1–2 (Hebrew) #

Fraenkel AH (1942) השימוש ב'דרישות' בבניין האריתמיקה והגאומטריה. קובץ הרצאות ומאמרים. הוצאת הסתדרות המורים בא"י. תש"ב. [The use of 'requirements' in developing arithmetics and geometry. Lectures and articles. The teachers' association in the land of Israel publishing]. (Hebrew)

Fraenkel AA (1942) מבוא למתמטיקה: בעיות ושיטות מן המתמטיקה החדשה. המספר והפונקציה. [Introduction to Mathematics: vol 1. Problems and methods of modern mathematics. Number and function]. Masada Publishing, Ramat Gan and then Hebrew University Press Association, Jerusalem. 381pp (Hebrew)

Fraenkel AH (1942) דוד ששון [David Sassoon]. *הצופה* Ha'Tzofe 6 .כ"ד באלול תש"ב Sept 1942 (Hebrew) #

Fraenkel AH (1942) על אמירת הסליחות. [The recitation of Selichot]. *הצופה* Ha'Tzofe ט' בתשרי תש"ב. [note a mistake in the Hebrew date]. 20 Sept 1942 (Hebrew) #

Fraenkel A (1943) *הילודה בישוב ובעיותיה*. [The birthrate in the Jewish population and its problems]. D. B. Aaronson, Jerusalem, 46pp. Series for current challenges (Hebrew)

Fraenkel AH (1943) על המבנה המתמטי של הלוח העברי מתוך דפים למתמטיקה ולפיסיקה לנוער המתלמד. טבת תש"ג חוברת ה'. [The mathematical structure of the Jewish calendar]. Dapim LeMatematica UlePhysica. [Pages in Mathematics and Physics for Youth]. Ch 1 and Ch 2. 5th booklet (Hebrew)

Fraenkel AH (1943) יצחק ניוטון 1642–1942: דברים שנאמרו על-ידי ד"ר י.ל. מאגנס, ..., א.ה. פרנקל... י.רקח... בחגיגת ניוטון שנערכה באוניברסיטה העברית בירושלים, ביום ג' אדר תש"ג [1943]. ירושלים : חברה להוצאת ספרים על-יד האוניברסיטה העברית Isaac Newton: 1642–1942; at the Isaac Newton celebration held at the Hebrew University in Jerusalem on the tenth of March 1943, University Press

Fraenkel AA (1943) Problems and methods in modern mathematics – 1. Scripta Math IX(1):81–84

Fraenkel AA (1943) Problems and methods in modern mathematics – 2. Scripta Math IX(2):162–168

Fraenkel AA (1943) Problems and methods in modern mathematics – 3. Scripta Math IX(3):245–255

Fraenkel AA (1943) Problems and methods in modern mathematics – 4. Scripta Math IX(4):5–18

Fraenkel AH (1943) העליה הפנימית בישוב. [Internal immigration in the y*ishuv*] גליון קעו. הד-ירושלים. Hed-Yerusahlayim. Issue 176. 1 Jan 1943 (Hebrew) #. Also reprint of unknown origin. pp 1–12. (Hebrew) ##

Fraenkel AH (1943) הזהרו בטעמי המקרא. [Pay attention to the leining marks] הארץ י"ב בשבט תש"ג Ha'aretz, 18 Jan 1943. (A response to this article is in *Ha'aretz* 31st January 1943 under the title 'עניינים' [issues]#)

Fraenkel AH (1944) Introduction in: Adult education: extension lectures and courses of the Hebrew University, Jerusalem 1943–1944. Achva Coop. Printing Press, Jerusalem, 19pp (Copy exists in Bivliotheek Universiteit van Amsterdam. Item 0150000448)

Fraenkel AH (1944) [Manners in the land]. B'nai Brit newsletter, 5. נמוסים בישוב. מתוך דברים שנאמרו ע"י האח פרופ' א"ה פרנקל באספת לשכת "ירושלים" סיון תש"ד. תדפיס מיוחד מתוך "ידיעות בני ברית" חוברת ה', תמוז תש"ד. יוצא לאור ע"י לשכת הגליל הגדולה מס' י"ד, ירושלים אב תש"ד. (Hebrew)

Fraenkel AH (1944) נוסחה המקשרת את הלוח העברי בלוח המושלימי. מתוך דפים למתמטיקה ולפיסיקה לנוער המתלמד.חוברת י' אייר תש"ד. [A formula relating the Jewish calendar to the Moslem calendar]. Dapim LeMatematica UlePhysica. [Pages in Mathematics and Physics for Youth]. (Hebrew) 4–6

Fraenkel AH (1944) The University and adult education. Davar, 10 Feb 1944. (Hebrew)

Fraenkel AH (1944) הרב ד"ר חנוך הכהן אהרנטריי ז"ל [Rabbi Dr. Chanoch HaCohen Ehrentreu]. *הצופה* Ha'Tzofe, 5 Mar 1944, p 3 (Hebrew) #

Fraenkel, A.H. (1944) לבעית השבת [On the problem of the Sabbath.] דבר *Davar* 26th May 1944. (Hebrew) #

Fraenkel AH (1944) לדמותו של ברל כצנלסון ז"ל [A eulogy for Berl Katznelson]. *הצופה*. Ha'Tzofe, 18 Aug 1944 (Hebrew) #

Fraenkel AA (1944) Problems and methods in modern mathematics – 5. Scripta Math X(3–4)

Fraenkel AH (1945) "על נימוקיהם של דחיית אד"ו ושל הסדר גו"ח אדז"ט לשנים המעוברות". [On the reasons for delaying Rosh HaShana on Sundays, Wednesdays and Fridays, and on the order of leap years]. In Fishman YL (ed) *זכרון לנשמת הרב אברהם יצחק הכהן קוק למלאות עשר שנים לפטירתו, קובץ תורני-מדעי*. Jerusalem Mossad HaRav Kook, pp 180–185 (Hebrew)

Fraenkel AA (1945) Introduction, revision and annotation of "פילוסופיה של המתמטיקה." "Philosophy of mathematics (Hebrew)" by Weyl H. Translated by D. Meuhad (Klinghofer). The Hebrew University Press Association, Jerusalem

Fraenkel AH (1945). בין ב. כצנלסון והאוניברסיטה העברית. Berl Katznelson and the Hebrew University. Davar, 3 *הדבר כ"ד באב תש"ה* Aug 1945 (Hebrew) #

Fraenkel AH (1945). ההכשרה הגופנית והחינוך לנימוסים. תמצית הרצאתו של הפרופ' אברהם הלוי פרנקל. Physical training and an education in manners. A summary of Prof. Fraenkel's lecture. *החינוך הגופני* Physical Education (4):3–4 (Hebrew) #

Fraenkel AH (1945) יחס הנוער שלנו אל התנ"ך [Our youth's attitudes towards the Bible]. In a section מדור לשאלות חנוך [Section on educational matters] *ידיעות בני ברית* אב תש"ה B'nai Brit News, pp 18–20 (Hebrew) #

Fraenkel AH (1946) Address by Abraham A. Fraenkel. In: Founder's Day: The Dropsie College for Hebrew and Cognate learning. Addresses: The Honorable Herbert H. Lehman, Professor Abraham Fraenkel. Dropsie College, Philadelphia, PA, pp 9–12 (English)

Fraenkel AA (1946) Einleitung in die Mengenlehre. Dover Publications, New York, NY, 424p

Fraenkel AA (1947) The recent controversies about the foundations of mathematics. Scripta Math XIII(4):17–36

Fraenkel AH (1947) The Hebrew University and the regulation of secondary education in Palestine. Jewish Education 18:2

Fraenkel AH (1947) .השוואות רחוקות מן המציאות. [Far-fetched comparisons]. A letter to the editor. Ha'aretz, 9 Nov 1947 (Hebrew) #

Fraenkel AH (1948) The relation of equality in deductive systems. Library of the Xth Intern. Congr. of Philosophy, Proceedings of the Congress, vol I, Amsterdam, pp 752–755 (published 1949)

Fraenkel A (1948) Principia Mathematica. In the section titled, 'Readers' Letters'. Letter to the editor. Palestine Post, 27 Jan 1948 (English) #

Fraenkel AH (1948). זרמים חדשים בפילוסופיה. רשמים מן הקונגרס הבינלאומי העשירי לפילוסופיה. [New currents in Philosophy. Impressions from the 10th International Congress on Philosophy]. *הדור*. Ha'Dor, חשון תש"ט Nov 1948 (Hebrew) #

Fraenkel AH (1948) על יסודות הלוח העברי. שיחה ראשונה. [The basis of the Hebrew Calendar. First conversation]. *הצופה לילדים* כרך ב' חוברת 32-33 , י"ג בניסן תש"ח, 22.4.48 עמוד 392. Ha'Tzofe for Children, vol 2, issue 32–33, 22 Apr 1948, p 392 (Hebrew)

Fraenkel AH (1948) .על יסודות הלוח העברי. שיחה שנייה [The basis of the Hebrew Calendar. Second conversation]. הצופה לילדים כרך ג', חוברת 1, תש"ט 3.9.1948 עמוד 15-14 Ha'Tzofe for Children, vol 3, issue 1, 3 Sept 1948, pp 14–15 (Hebrew)
Fraenkel AH (1949) .על יסודות הלוח העברי. שיחה שלישית [The basis of the Hebrew Calendar. Third conversation]. *הצופה לילדים* Ha'Tzofe for Children, 27 Jan 1949, pp 204–205 (Hebrew) ##
Fraenkel AH (1949) Problems and methods in modern mathematics. Scripta Math XV:169–182
Fraenkel AH (1949) הרב שמחה אסף – בן 60 [Rav Simcha Assaf – at age 60]. *Ha'aretz* תש"ט. א' בתמוז 28 June 1949 (Hebrew) #
Fraenkel AH (1949) התפילות והפיוטים בארצנו ובזמננו [The prayers and the liturgical poems (*piyyutim*) in our land and in our time]. *הצופה* Ha'Tzofe. י"ד בניסן תש"ט 13 Apr 1949 (Hebrew) #
Fraenkel AH (1949) מסדר "בני ברית" בישראל [B'nai Brit Order in Israel]. דבר Davar י"ב בתשרי תש"י 5 Oct 1949)Hebrew) #
Fraenkel AH (1949) פכים קטנים [Notes]. *הדור* Ha'Dor י' בטבת תש"ט 11 Jan 1949 (Hebrew) #
Fraenkel AH (1949–1968) Editor for mathematics for the Encyclopedia Hebraica and author of multiple entries, vols 1–19. Encyclopedia Publishing Co., Jerusalem (Hebrew)
Fraenkel AH (1950) האינסוף במאתימאטיקה [Infinity in mathematics]. Entry in the Encyclopedia Hebraica, vol 2, pp 959–960. Encyclopedia Publishing Co., Jerusalem (Hebrew)
Fraenkel AH (1950) "נופה ואקלימה של ארץ ישראל כמדריך להבנת ספרותו העתיקה" ["The Landscape and Climate of Israel as a Guide to Understanding its Ancient Literature"]. In: Asaf S, Even Shmuel Y, Benjamin R, (eds) "מנחה ליהודה" ["A present to Yehudah"], Jerusalem, Rabbi Kook Institute and the Hebrew Association in South Africa (Hebrew)
Fraenkel AH (1950) מאורעות בלוח העברי [Events in the Hebrew Calendar]. Unknown journal, around end of March or beginning of April 1950 (Hebrew) #
Fraenkel AH (1950) "פורים משולש" בירושלים ושבת ערב פסח בכל העולם [A three day Purim in Jerusalem and a Sabbath on Erev Passover in the whole world]. *ההד* אדר תש"י עמודים כז-כח Ha'Hed, March 1950, pp 27–28 (Hebrew) ##
Fraenkel AH (1950) Influencing secondary education. In: Spiegel M (ed) The Hebrew University of Jerusalem 1925–1950, Semi-Jubilee Volume, 1950. Goldberg Press, Jerusalem, pp 108–110. (English) Note: It seems like the following chapter, entitled 'Adult Education' (pp 111–114) was also written by Fraenkel, but the authorship is not given
Fraenkel AH (1950) The commission for secondary education. In: Toren H (ed) The Hebrew University of Jerusalem 1925–1950 א. ה. פרנקל. הוועדה לבתי ספר תיכוניים. פרק בספר: האוניברסיטה העברית בירושלים - כה שנה תרפ"ה – תש"י, עורך: חיים תורן עמ' 83–85, ירושלים תש"י. (Hebrew) Note: It seems like the chapter entitled 'Adult Education' pp 97–99 was also written by Fraenkel, but the authorship is not given
Fraenkel AH (1951) On the crisis of the principle of excluded middle. Scripta Math XV(1–2):5–16
Fraenkel AA (1952) L'axiome du Choix. Revue philosophique de Louvain 50(27):439–459 (Cf. *L'Enseignement Math.* 34 (1935), 32–51)
Fraenkel AH (1952) .חיים וייצמן ומדעי הטבע באוניברסיטה העברית [Chaim Weizmann and the sciences at the Hebrew University.] דבר Davar, 10 Dec 1952 (Hebrew) ##
Fraenkel AH (1952) דבר הנשיא [Word of the President]. .בני ברית לשכת הגליל הגדולה לישראל מס' י"ד .עלון מודיעין Alon Modiin of B'nai Brit Large Lodge number 14, vol 6, May 1952 (Hebrew) ##
Fraenkel A (1953) *.מבוא למתמטיקה: בעיות ושיטות מן המתמטיקה החדישה: אריתמטיקה ואלגברה* [Introduction to Mathematics: Problems and methods of modern mathematics: arithmetic and algebra]. Masada Publishing, Ramat Gan, Book (Hebrew)
Fraenkel AA (1953) Abstract Set Theory, 1st ed, Studies in logic and the foundations of mathematics. North-Holland, Amsterdam, 479pp
Fraenkel AH (1953) Archimedes. Entry in the Encyclopedia Hebraica, vol 5, pp 926–929. Encyclopedia Publishing Co., Jerusalem (Hebrew)
Fraenkel AH (1953) With tribulations, signs and wonders, and in war and in great terror. במסות באותות ובמופתים ובמלחמה ובמוראים גדולים. *המעיין* א, ב (תשרי תשי"ד) HaMaayan 1(2):7–11 (Hebrew)

Fraenkel A (1953) .מבוא למתמטיקה: בעיות ושיטות מן המתמטיקה החדישה: האינסוף והמרחב. תורת הקבוצות [Introduction to Mathematics: Problems and methods of modern mathematics, vol 2. Infinity and space. Set theory]. Magnes Press/Masada Publishing, Jerusalem/Ramat Gan. Book. 463pp (Hebrew)

Fraenkel AH (1953) על ימים ושנים. [On days and years]. דבר Davar, 23 Oct 1953 (Hebrew) ##

Fraenkel A (1954) The intuitionistic revolution in mathematics and logic. J Bull Res Council Israel 3:283–289

Fraenkel AH (1954) "התלכדות בין פסח לפסחא של הנוצרים" [The convergence of passover with the Christian holiday of Easter] דבר Davar, 16 Apr 1954 (Hebrew) ##

Fraenkel AH (1954) Issues of continuing education among workers. In: Banaari N (ed) Orot – On questions of culture and education. Tel-Aviv, pp 84–87. בעיות ההשכלה בהתישבות העובדת. אורות – לשאלות התרבות והחינוך קובץ ב', עורך נחום בנארי, תל – אביב 1954, עמ' 84 – 87 (Hebrew)

Fraenkel AH (1954) פורים משולש וערב פסח בשבת [A three day Purim and Erev Passover on the Sabbath]. מחברת בטאון חברת כיח לעינייני חנוך ותרבות, ספרות ומחקר. Les Cahiers de l'Alliance Israélite Universelle. Issue 1–2, year 3, April 1954, pp 14–15 (Hebrew) ##

Fraenkel AA (1954) Pesach and Easter coincidence. The Jewish Chronicle, 9 Apr 1954, p 21 (English) ##

Fraenkel AH (1955)."על סדר התפילות בקיבוץ הדתי" [On the order of prayers in the religious kibbutz] In: Tirosh Y (ed) שי לישעיה: ספר יובל לר' ישעיהו וולפסברג בן הששים Merkaz LeTarbut Shel HaPoel Mizrahi, Tel Aviv, pp 193–194

Fraenkel AA (1955) Integers and theory of numbers. Problems and methods in modern mathematics. The Scripta Mathematica Studies Number 5 (dedicated to the memory of Chaim Weizmann). Mack Printing, Easton, PA. Book. 102pp

Fraenkel AH (1955) Topography of Eretz Israel, vol 21, no 5. The Synagogue Light, New York, NY

Fraenkel AA (1955) .מבוא למתמטיקה: בעיות ושיטות מן המתמטיקה החדשה. אנליזה [Introduction to Mathematics: Problems and methods of modern mathematics. The number and the function, analysis]. Masada Publishing, Ramat Gan and then Hebrew University Press Association, Jerusalem (Hebrew)

Fraenkel AH (1955) Geometry. Entry in the Encyclopedia Hebraica, vol 10. Encyclopedia Publishing Co., Jerusalem, pp 106–109 (Hebrew)

Fraenkel AH (1955) Non-Euclidian Geometry. Entry in the Encyclopedia Hebraica, vol 10. Encyclopedia Publishing Co., Jerusalem, pp 118–121 (Hebrew)

Fraenkel AH (1955) Synthetic geometry. Entry in the Encyclopedia Hebraica, vol 10. Encyclopedia Publishing Co., Jerusalem, pp 121–122 (Hebrew)

Fraenkel AH (1955) Projective and affine geometry. Entry in the Encyclopedia Hebraica, vol 10. Encyclopedia Publishing Co., Jerusalem, pp 122–126 (Hebrew)

Fraenkel AH (1955) על המחקר במדעי הטבע באוניברסיטה. ידיעות האיניברסיטה העברית בירושלים. כרך 5 חוברת א'. עמוד 3. [On research in the sciences at the university]. Bulletin of the Hebrew University 5(1):3 (Hebrew)

Fraenkel AH (1955) Multi-dimensional geometry. Entry in the Encyclopedia Hebraica, vol 10. Encyclopedia Publishing Co., Jerusalem, pp 126–127 (Hebrew)

Fraenkel AH (1955) Gauss, Carl Friedrich. Entry in the Encyclopedia Hebraica, vol 10. Encyclopedia Publishing Co., Jerusalem, pp 143–148 (Hebrew)

Fraenkel AH (1955) Galois Évariste. Entry in the Encyclopedia Hebraica, vol 10. Encyclopedia Publishing Co., Jerusalem, pp 788–789 (Hebrew)

Fraenkel AH (1955) החינוך התיכון. (מכתב) הארץ 10.3.1955. [Secondary education]. (Letter) Ha'aretz, 10 Mar 1955 (Hebrew)

Fraenkel AH (1955) אלברט איינשטיין. משא כ"א באייר תשט"ו. [Albert Einstein] Masa, 13 May 1955 (Hebrew)

Fraenkel AA (1956) Review: Franz von Krbek, Wohlordnung. J Symb Logic 21(1):96–97

Fraenkel AH (1956) מבעיות המגמות בחינוך התיכון. בחינוך ובתרבות. משרד החינוך והתרבות. גליון 6 יוני. עמודים 9-10. וכן נדפס כ 'הערות לויכוח', בתלמים ירחון תנועת המושבים. חוברת קע"ו. תל-אביב. [Difficulties in the trends of secondary education.] Education and Culture. Published by the Ministry of Education and Culture. June, pp 9–10. Also published in Telamim, the monthly *Journal of the Moshav Movement*, #176. Publication of the Moshavim movement, Tel-Aviv. (This article is followed by one by M. Ziv which discusses Fraenkel's points of view regarding secondary education - pp 10–12) (Hebrew)

Fraenkel AH (1956) יעקב לויצקי. הארץ. Jacob Levitzki. Ha'aretz, 28 Mar 1956

Fraenkel AH (1956) המחקר בישראל ותנאיו החמריים. בטרם. 1.2.56. [Research in Israel and its practical conditions.] Be'Terem, 1 Feb 1956. A similar article was published in Ha'aretz, 20 Jan 1956. (Hebrew)

Fraenkel AH (1956) יעקב לויצקי ז"ל Jacob Levitzki. גליונות מתמטיקה ופיסיקה. Gilyonot Matematika UPhisica, vol 8. Apr 1956, pp 110–111

Fraenkel A (1954/1957) [two volumes]. *מבוא למתמטיקה: בעיות ושיטות מן המתמטיקה החדישה: האינסוף והמרחב. גאומטריה.* [Introduction to Mathematics: Problems and methods of modern mathematics: Infinity and space. Geometry]. Masada Publishing, Ramat Gan

Fraenkel AA, Bar-Hillel Y (1958) Foundations of Set Theory. Studies in Logic and the Foundations of Mathematics. North-Holland, Amsterdam. Book. 415pp

Fraenkel A (1958) Historical introduction. In: Bernays P (ed) Axiomatic Set Theory. Studies in logic and the foundations of mathematics series 21: 1, pp 1–35. North-Holland, Amsterdam

Fraenkel AH (1958) "משום ירקיא - משום מתיא". [Because of the vegetables – because of the dead] (explanations for determining changes of date in the Jewish calendar). In: Braunstein S, Chorgon G (eds) *ספר יובל לכבוד שמואל קלמן מירסקי*. Vaad HaYovel, New York, NY, pp 248–250 (Hebrew)

Fraenkel A (1958) Paul Bernays und die Begründung der Mengenlehre. Dialectica Int J Philos 12:274–279

Fraenkel AA (1959) Mengenlehre und Logik, Berlin (Erfahrung und Denken, vol 2). Duncker & Humblot. Book. 110pp

Fraenkel AA (1959) Philosophie der Mathematik. In: Die Philosophie im XX. Jahrhundert, herausg. v. *F. Heinemann* (Stuttgart), pp 334–359

Fraenkel AA (1959) Review: Katuzi Ono, A set theory founded on unique generating principle. J Symb Logic 24(3):227

Fraenkel AH (1959) פסח ופסחא. [Passover and Easter]. *הארץ* Ha'aretz, 3 June 1959 (Hebrew) ##

Fraenkel AH (1959) על נוסח הסליחות. מכתב למערכת. [On the rendition of the Selichot. Letter to the editor]. עמודים Amudim אלול Sept 1959, pp 7–8 (Hebrew) ##

Fraenkel AA (1960) Epistemology and logic. Synthese 12(4):333–337

Fraenkel A (1960) Theory of sets. In: Encyclopedia Britannica

Fraenkel AA (1960) Jewish mathematics & astronomy. Scripta Mathematica: A Quarterly Journal Devoted to the Philosophy, History, and Expository Treatment of Mathematics 25:33–47. (English) (Appeared in Hebrew in *Tekhnika Umada series 65*, Tel Aviv, 1947. 30pp. 'המחקר המתמטי והאסטרונומי אצל היהודים. תל-אביב, 'טכניקה ומדע. Footnote 12 of the *Scripta Mathematica* version claims that the essay was written in the 1930s)

Fraenkel AA (1961) Abstract Set Theory, 2nd completely revised edition. Studies in logic and the foundations of mathematics. North-Holland, Amsterdam. Book. 295pp

Fraenkel AA (1962) A biography of G. Cantor. In: Zermelo E (ed) Gesammelte Abhandlungen mathematischen und philosophischen Inhalts. Mit erläuternden Anmerkungen sowie mit Ergänzungen aus dem Briefwechsel Cantor-Dedekind. G. Olms Publishing, Hildesheim

Fraenkel AA (1964) Extension of the number-concept: groups and fields, rational, real, complex, hypercomplex numbers. The Scripta Mathematica Series Number 10. Yeshiva University, New York, NY, 88p

Fraenkel AH (1964) "עקרון הבחירה ובעית הרצף (הקונטינואום)." [The principle of choice and the continuum problem.]. דבר האקדמיה הלאומית הישראלית למדעים, כרך ראשון חוברת 2. Israel National

Academy of Sciences, 1(2) Magnes Press, Hebrew University. Central Press, Jerusalem, pp 1–10. דפוס מרכז (Hebrew)
Fraenkel AA (1966) Abstract Set Theory, 3rd revised edition. North-Holland, Amsterdam, 297pp
Fraenkel AA (1966) Logik und Mathematik. Studium generale 19:127–135
Fraenkel AA (1966) Set Theory and Logic. Addison-Wesley, Reading, MA. Book. 102pp
Fraenkel AH (1966) "עיבור השנים וקידוש החודש" [Intercalation of years and sanctification of the month]. In: *אמונה, דת ומדע*. Misrad HaChinuch VeHaTarbut, Jerusalem, pp 61–70 (Hebrew)
Fraenkel AA, Bar-Hillel Y (1966) Foundations of Set Theory, 2nd revised edition with A. Levy as co-author. North-Holland, Amsterdam
Fraenkel AA, Bar-Hillel Y (1966) Foundations of Set Theory. Osnovaniia teorii mnozhestv. Mir Publishing, Moskva
Fraenkel AA (1966) Das Leben Georg Cantors. In: Cantor, Georg: Gesammelte Abhandlungen mathematischen und philosophischen Inhalts/Georg Cantor. Mit erl. Anm. sowie mit Ergänzungen aus dem Briefwechsel Cantor - Dedekind. Hrsg. von Ernst Zermelo. Nebst einem Lebenslauf Cantors von Adolf Fraenkel. - Reprografischer Nachdr. d. Ausg. Berlin 1932. - Hildesheim [u.a.]: Olms, 1966. – pp 452–483. (Olms-Paperbacks ; 24) [Sammlung]. - Reprografischer Nachdr. d. Ausg. Berlin 1932
Fraenkel AA (1967) Lebenskreise: aus den Erinnerungen eines judischen Mathematikers. Deutsche Verlags-Anstalt, Stuttgart, 207p
Fraenkel A (1967) The notion of 'definite' and the independence of the axiom of choice (1922). Reprinted in van Heijenoort, From Frege to Gödel: a source book in mathematical logic, 1879–1931, Harvard University Press
Fraenkel AA (1967) The principle of choice and the continuum problem. The Proceedings of the Israel Academy of Sciences and Humanities, vol 1, Chap. 2. Central Press, Jerusalem (English)
Fraenkel AA (1968) Mengenlehre und Logik. Duncker & Humblot, Berlin, 110pp
Fraenkel AA (1968) Historical introduction. In: Bernays P (ed) Axiomatic Set Theory, 2nd ed. Studies in logic and the foundations of mathematics. North-Holland, Amsterdam
Fraenkel AA (1968) Abstract Set Theory. 4. VIII. North-Holland, Amsterdam [u.a.], 297 S (Studies in logic and the foundations of mathematics)
Fraenkel AH (1969) לוח Calendar. Entry in the Encyclopedia Hebraica, vol 21. Encyclopedia Publishing Co., Jerusalem, pp 335–353 (Hebrew)
Fraenkel AA (1970) Teoria degli insiemi e logica. Translated by Marfurt M. Ubaldini Editore, Roma, 108pp (Italian)
Fraenkel AA (1971) Teoría abstracta de conjuntos. Tarea, Revista de iniciación filosófica. Universidad Nacional de La Plata, Pertence, 137pp
Fraenkel AA (1972) Zehn Vorlesungen über die Grundlegung der Mengenlehre. Wissenschaftliche Buchgesellschaft, Darmstadt. Book. 182pp
Fraenkel AA (1972) Einleitung in die Mengenlehre. Sändig publisher, Walluf (bei Wiesbaden), 424pp, 13 graph (German)
Fraenkel AA, Bar-Hillel Y, Levy A with the collaboration of Dirk van Dalen (1973) Foundations of Set Theory, 2nd revised edition. Studies in logic and the foundations of mathematics, vol 67. North-Holland, Amsterdam, 404pp
Fraenkel AA (1976) Teoría de los conjuntos y lógica. Universidad Nacional Autónoma, Mexico. Instituto de Investigaciones Filosóficas, 136pp (Spanish)
Fraenkel AA, Lévy A (1976) Abstract Set Theory, 4th revised edition. North-Holland, Amsterdam
Fraenkel AA, Bar-Hillel Y, Levy A (1984) Foundations of Set Theory, 2nd revised edition. Studies in logic and the foundations of mathematics, vol 67. North-Holland, Amsterdam
Fraenkel AA (1991) Historical introduction. In: Bernays P (ed) Axiomatic Set Theory. Dover Publications, New York, NY
Fraenkel AA (2004) Integers and theory of numbers, Dover Phoenix edition. Dover Publications, Mineola, NY, 102pp
Fraenkel AH (unknown) בעיית המקבילים והקשר בינה לבין שאלת האינסוף. [The problem of parallel lines and its relation to the question of infinity]. Unknown journal, pp 2–3 (Hebrew) #

Fraenkel AH (unknown) האיגוד לקידום המדע בישראל [The Israel Society for the Advancement of Science]. Unknown journal (Hebrew) ##

Fraenkel AH (unknown) פורים ושושן בשבת. [Purim and Shushan Purim on the Sabbath]. *הארץ* Ha'aretz (Hebrew) ##

Fraenkel AH (unknown) Über die praktische Arbeit des deutschen Misrachi in Eretz-Israel, 2p (German) #

Fraenkel AH (unknown) חקירת המצוות – מצווה [Investigation of "The Commandments" is a commandment]. Unknown Journal 9–10. (Hebrew)

Published Interviews with Fraenkel

(1936) 500 Studenter ved Jerusalems Universitet [500 students at Jerusalem University] Studentenes ydelser gjennemgående høiere enn i Europa.[Students' performance better than in Europe] En samtale med den palestinske regjerings representant ved matematiker-kongressen, Professor Fraenkel. [A conversation with the Palestinian government representative at the Mathematical Congress, Professor Fraenkel.] Tidens Tegn tirsdag, 14 July 1936 Oslo 9 (Norwegian) #

(1951). התיכון בישראל [Secondary education in Israel]. *דבר השבוע*. Dvar HaShavua 9:5. 1 Mar 1951 (Hebrew) #

(1955) על הרבנות הראשית לישראל. *פנים אל פנים*. י"א, אדר תשט"ו. [On the Chief Rabbinate in Israel]. Panim el Panim 11:12. 2 Mar 1955 (Hebrew) #

(1960) תלמיד מיוחד במינו [A unique student]. In the section כשהפרופסור היה סטודנט [When the Professor was a student] *פי האתון* Pi Ha'Aton, 25 May 1960 (Hebrew) ##

(1961) מדוע הצטרפתי למפלגה הליברלית? Why did I join the liberal party? Ha'Boker, 26 May 1961, p 3 (Hebrew) #

(1961) כשהפרופסור לא היה פרופסור [When the Professor had not been a professor] Pi Ha'Aton, 6 Dec 1961 (Hebrew) ##

Index of Names

A.A. Fraenkel, *Recollections of a Jewish Mathematician in Germany*,
DOI 10.1007/978-3-319-30847-0

C

D

G

O

P

R

The manufacturer's authorised representative in the EU is Springer Nature Customer Service Centre GmbH, Europaplatz 3, 69115 Heidelberg, Germany. If you have any concerns regarding our products, please contact ProductSafety@springernature.com

Printed and bound by CPI Group (UK) Ltd, Croydon, CR0 4YY
30/09/2026
02225454-0003